AF870983

Autodesk Inventor Certified User Exam Study Guide

Inventor 2027 Edition

L. Scott Hansen, Ph.D.

SDC PUBLICATIONS

SDC Publications
P.O. Box 1334
Mission, KS 66222
913-262-2664
www.SDCpublications.com
Publisher: Stephen Schroff

Copyright 2026 Scott Hansen, Ph.D.

All rights reserved. This document may not be copied, photocopied, reproduced, transmitted, or translated in any form or for any purpose without the express written consent of the publisher, SDC Publications.

It is a violation of United States copyright laws to make copies in any form or media of the contents of this book for commercial or educational purposes without written permission.

Examination Copies
Books received as examination copies are for review purposes only and may not be made available for student use. Resale of examination copies is prohibited.

Electronic Files
Any electronic files associated with this book are licensed to the original user only. These files may not be transferred to any other party.

Trademarks
The following are registered trademarks of Autodesk, Inc.: Autodesk Inventor, 3D Studio, ADI, Advanced Modeling Extension, AME, AutoCAD, AutoCAD Mechanical Desktop, AutoCAD Development System, AutoCAD LT, Autodesk, Autodesk Animator, AutoLISP, AutoShade, AutoVision, and Heidi. The following are trademarks of Autodesk, Inc.: ACAD, Autodesk Device Interface, AutoCAD DesignCenter, AutoTrack, Heads-up Design, ObjectARX and Visual LISP. Microsoft and Windows are either registered trademarks or trademarks of Microsoft Corporation. All other trademarks are trademarks of their respective holders.

The author and publisher of this book have used their best efforts in preparing this book. These efforts include the development, research and testing of the material presented. The author and publisher shall not be liable in any event for incidental or consequential damages with, or arising out of, the furnishing, performance, or use of the material.

ISBN-13: 978-1-63057-835-0
ISBN-10: 1-63057-835-5

Printed and bound in the United States of America.

Table of Contents

Chapter 1:
Potential Value of Certification

Certification is a tremendous achievement that is sought by many people for several activities and skills. Autodesk offers professional and user level certification exams through their certification partner.

Autodesk® Inventor™ offers an enormous number of features, but not all of them are included in the certification exams. This guide will focus on the objectives for the User level certification for Autodesk Inventor.

Why certify?

For many who've taken the Autodesk Inventor Certified User exam, it's an opportunity to validate their skills with a challenging design tool. In some other cases, it's required as a mark to show the completion of a course or to meet a mandate.

There are many reasons to pursue certification. In a competitive workplace or job market, people will use certification to differentiate themselves from other people with similar experience and work history.

If you seek employment with a company that's using Inventor, being certified as a User or Professional can help you become more productive at a faster rate and possibly more likely to win the job.

Even if you're not familiar with what other companies use in the job market, having certification can demonstrate a high skill level to potential employers. This can also be true for employers that use a tool competitive with the one in which you are certified. Many of the direct competitors of Autodesk Inventor have very similar workflows. Having a high degree of confidence in those workflows can shorten the learning curve for adopting the new tool.

Though certification is popular, only a small minority of users hold certification. By successfully gaining certification, you will join a select group of professionals. Attaining certification might appear even more impressive when achieved as a student. User certification in particular can be achieved in a relatively short time with concerted effort and a strong desire to pass the test.

The secure process for taking the exam is also appealing. It is one thing to be able to speak to proficiency in the tool, but an objective test taken in a specific environment in limited time is a much stronger statement of skill.

An often-overlooked benefit of preparing for Certification is that it can also be used as an effective way to learn a broad variety of the tools in the software. By using the certification objectives as a guide for studying, you'll be exposed to more of the software than you already are in your daily work or class work assignments.

What do you receive?

Once you have passed the certification exam, you will be able to download a printable certificate and will receive an email notification that you have a badge available to display on social media or to use as a logo on certain professional materials such as business cards.

The certificate will have a unique identifying number that can be used to validate your certification. It will be printed along with your name and the date of the certification. The certificate will also list the product and the level in which you are certified.

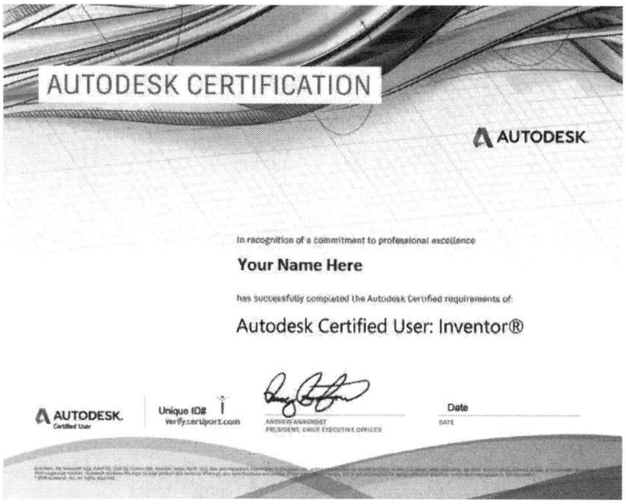

Chapter 2:
Preparing to Take the Exam

The process of preparing for certification begins with you. While there are tools and materials available to help you prepare, it is ultimately your decision to pursue certification and set everything in motion.

When will you be ready?

It is really up to you to decide when you're ready to take the certification exam. There are general guidelines that can be found, but this chapter will focus on offering you information and suggestions to help you make the decision on when you're prepared.

How much experience do I need?

You can find recommendations online that suggest a minimum of 150 hours experience using Autodesk Inventor before attempting User certification. While this is not a bad suggestion, the reality is far more subjective. It's important to keep the suggested 150 hours in perspective. Using a select group of tools repeatedly for 150 hours is not an effective way to prepare for the exam. You need to spend time in each of the objective domains that are in the exam.

Using the certification objectives as a reference, review them carefully and see how many topics you recognize as tools or workflows that you've used and are comfortable with. If you find certification objectives that you're not familiar with, you can begin looking for learning materials that address these topics.

In the following chapters of this book, descriptions are offered on the certification objectives along with explanations on how and why the tools are used. This is intended to help you understand the value of learning a tool that you might not use in your normal day-to-day or classroom work, but there is no step-by-step learning practice offered on the objectives.

User Certification objectives

The certification objectives for this Autodesk Inventor User exam focus on the tools that are commonly used to support the core workflows of Autodesk Inventor. These objectives are organized into groups or domains that make it easier to sort through when reviewing.

The table below will show the objective groups and the topics within them. Use this as a quick reference to find the topics that you're less familiar with if you have limited time to review all of them.

USER INTERFACE AND NAVIGATION	
	Change the viewpoint using the ViewCube
	Change setting of the ViewCube
	Understand Inventor file types and standard templates
SKETCHING	
	Apply dimensions to a sketch
	Assign geometric constraints
	Project geometry
	Create and modify geometric shapes
	Modify an Inventor model
PART MODELING	
	Create extrude features
	Create a pattern of features
	Create a shell feature
	Apply fillets and chamfers
	Create hole features
	Create revolve features
	Place threads

BROWSER EDITING	
	Suppress and un-suppress part features
	Toggle visibility of features and sketches
ASSEMBLY MODELING	
	Apply basic assembly constraints (mate, flush, insert, directed angle)
	Ground base component of an assembly
	Apply an offset to constrained parts
	Determine the degrees of freedom of a component
	Create a presentation model
DRAWING	
	Control sheet size and add a title block
	Select and place a front view
	Create a drawing view from an existing view
	Add annotation and dimensioning to a drawing
	Add sheets to a drawing
	Create a drawing view based on an assembly and presentation file
	Add balloons to a drawing

User versus Professional

An easy way to frame the differences between the User and Professional certification exam objectives is extensibility. The professional exam includes objectives on subjects like weldments, advanced shaped description, and advanced assembly functions. This is why the professional exam recommendation for experience is a minimum of 400 hours, and once again, these are hours spent across the variety of tools related to the objectives.

It is not necessary to take the User exam prior to taking the Professional exam. However, people new to certification and relatively new to Autodesk Inventor should find the user exam a challenge and a great way to become accustomed to the certification process. With User certification achieved, they can continue to develop their skills and add Professional certification when they're ready.

Is there a way to practice?

There are two different ways to prepare for the exam. The first is to get training on the tools addressed by the objectives or work through self-paced learning materials on the same. This will cover the Inventor skills, but many people find the exam itself to require a level of skill. Reading the exam problem and being able to mentally formulate a process to solve it is not a natural skill for many people. Practicing this can be useful if you're uncomfortable taking timed exams. Besides taking the actual exam, there are a couple of different ways to practice taking the test.

Appendix of this book

The appendix of this book contains thirty questions similar to those in an actual exam that do not require using Inventor to find a solution. Try working through these without the use of the software as a reference the first time through and then compare your answers to see where additional work needs to be done to prepare. This will help you to visualize the tools and help train you to think about how Inventor works to solve the problem. If you're working with others also preparing to take the exam, you can use these questions to quiz one another as well.

Autodesk Authorized Training Center (ATC)

Authorized Training Centers offer in-person and online live training on Autodesk products. Some might offer training specific to certification exams, but even if they don't, selecting a few standard courses could still cover most if not all of the certification objectives.

To find an Authorized Training Center or Authorized Academic Partner near you, you can use the locator link on the program website, enter your location, distance you're willing to travel, industry, and product you want training on. The locator will generate a list of training centers with information on how to contact them.

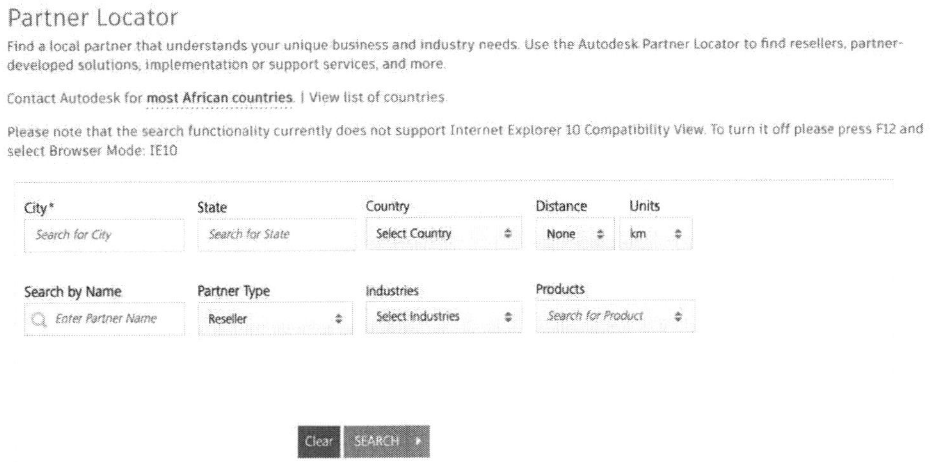

How do I take the exam?

Because the exam is an objective assessment of skill there is an elevated level of security around the contents of the exam and access to it. Presently the exam is only administered at a certification center and it requires the center to have a proctor to verify that external resources weren't used for the exam.

This protects the value of the exam and certification as a whole by making sure that people are certified based on their skill with Inventor and not their ability to use a search engine to find an answer.

Find a certification center

When you are ready to take the exam, the first step is usually finding a place to take it. At http://www.certiport.com/locator you can search for a certification center based on your location and the certification that you are pursuing. Some schools set themselves up as certification centers as well to be able to offer their students easier access to the exams.

If a center does not list specific times and dates for the exam you are looking for, contact them for information. A website, phone number or contact e-mail should be listed in the search results. Even if they do list times and dates, it is a good idea to contact the centers to make sure there is no other information that you need to be able to secure an exam time.

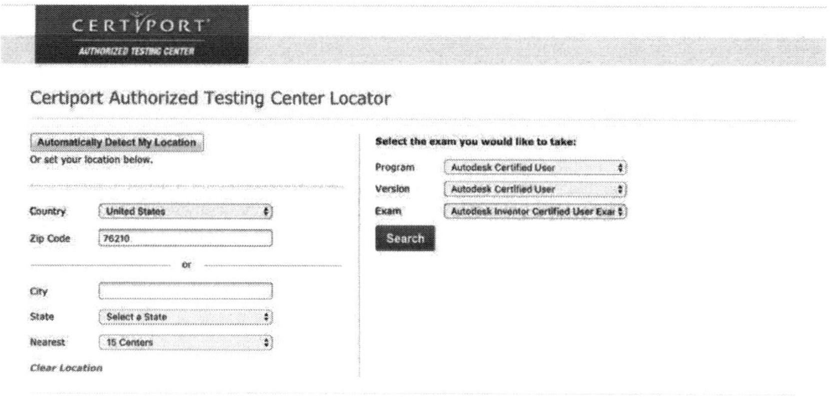

Purchase and register

Before taking the exam, you will need to have an account. The account can be created on its own or when purchasing a voucher through the shop.certiport.com website. Through the Certiport store you can currently purchase bundles that can include retakes or GMetrix practice exams. The Certiport account can be used for all future certifications including certifications outside of the Autodesk software tools. Pearson/Certiport offers certification for Microsoft, Adobe, and other products. If you have taken another certification exam through that company, you can use your existing login for accessing Autodesk certifications as well.

If you will be taking the exam through your school, they might be able to assist you with creating your account. If you choose to purchase a voucher you will have a number that entitles you to take the exam. The certification center can validate that number and use it to connect the exam with your account. The cost of the Autodesk Inventor User certification might vary depending on if you purchase it from the Certiport store. Schools offering the exam might offer a discount and process the voucher information for you.

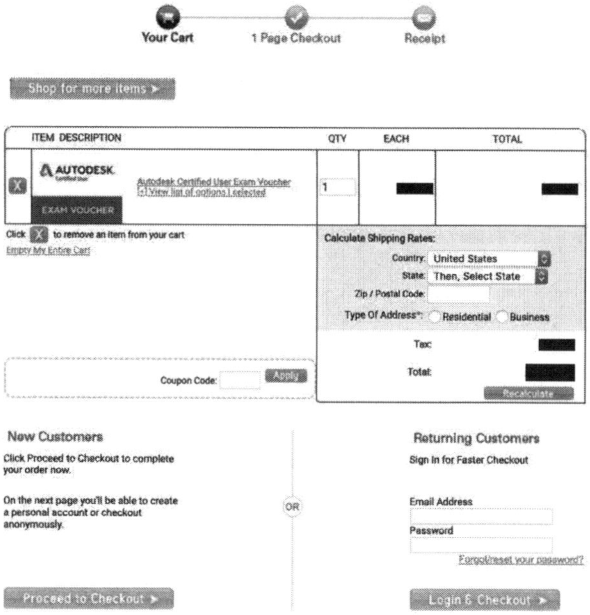

Tips for taking the exam

The Inventor User exam is thirty (30) questions and the exam time is a maximum of fifty (50) minutes. Before beginning the exam, carefully read all of the instructions regarding the tools used for the exam and any other requirements. The time spent reading instructions is not counted against your time to complete the exam.

The Autodesk Inventor User Certification exam is primarily questions that will ask you to perform a task or series of tasks in the software. You will not only need to know how to perform the task but also how to get the answer from the model or drawing.

Many of the exam questions will ask you for a volume of the model or the value of a center of gravity for the model once the work is completed.

Start the software before beginning the exam

This suggestion will be in the pre-exam instructions, but it is important enough to note separately. In order to use the software and complete the exam form you must be able to toggle between them. To do this, Inventor must be running before you begin the exam.

Once you're in the exam application you can use the Alt+Tab key to switch between them. This makes it very easy to quickly work between the programs.

iProperties dialog

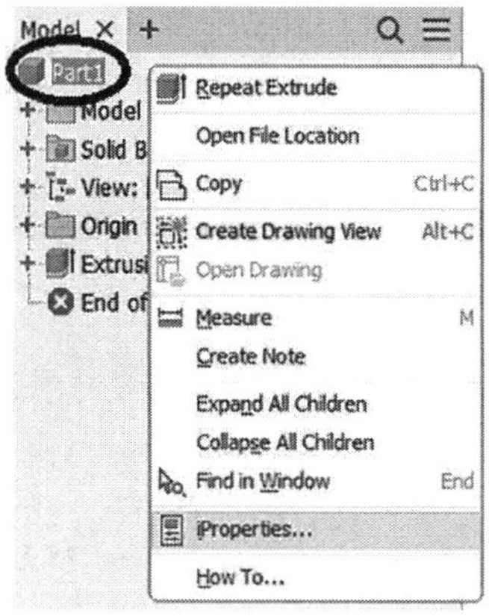

A frequent source for the answer you need is the iProperties dialog. It can be accessed by a right-click on the model icon at the top of the browser. The iProperties dialog has seven tabs. The one you will need information from is the Physical tab. The physical tab offers the Mass, Volume, and Center of Gravity information on the model. When you first enter the tab that data might not be displayed. Simply click the Update button to process the model.

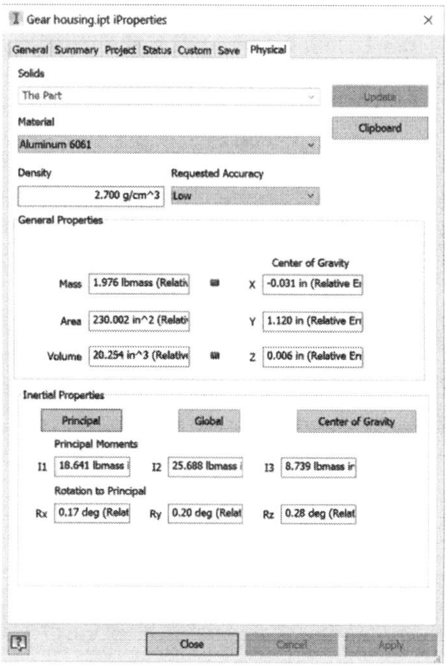

Note: It is critical that you do not change the precision level of the information as it might give you a result that the exam grading software does not recognize and mark the question as incorrect.

The answer the exam will be looking for is the entire number displayed in the dialog. If you choose to copy the result from the iProperties dialog and paste it into the answer area of the exam, be sure to copy all of the digits or letters displayed.

Be prepared to take notes

The Certification center should provide you with a sheet of paper and writing utensil. This can be very useful for results in the software that you cannot copy to the clipboard. You will not be able to remove any written notes from the room where the exam is being taken. This is another step to ensure the quality of the exam and its value.

Mark and return to difficult questions

This is true regardless of the exam's subject. When taking a timed exam it is best practice to pass over a question that you do not immediately recognize how to solve and move on to a question that you immediately know how to answer.

Button	Action
NEXT ▶	This button will move you forward to the next screen.
◀ BACK	This button will move you backward to the previous screen.
CALCULATOR ▦	This button will display a calculator. This button will be available to you once the test begins.
REVIEW ✓	This button will allow you to quickly review all questions.
HELP ?	This button will allow you to refer to the information contained in this tutorial at any point during your exam.
RESTART APP ↺	If for any reason the application has closed or is frozen you may restart it with this button. The test time will pause while the application is restarting.
SKIP INFO ❯❯	This button will allow you to skip the tutorial section.

The Inventor Certified User exam has the ability to mark a question for review. You can do this whether or not you select and answer or enter a value in the question. When you've successfully completed the questions that you recognized, you can have the exam software display what questions were marked and go back to them having preserved as much time as possible.

The exam tools can also display the number of questions and indicate whether an answer was given. If you forget to mark a question that you don't answer, this is a nice fall back so you don't miss out on what could be a correct answer.

Pause if you must

If you need a short break, it is possible to pause the exam. Pausing the exam will preserve where you are in the process and might require the proctor's assistance to resume.

In case of a crash

Inventor and the User Certification exam are computer programs and as such have some level of vulnerabilty to crashing. In the event the exam crashes, alert the proctor in the certification center as quickly as possible. The proctor will be able to restore your exam session and it should be recovered up to the last question answered.

If the Inventor software should happen to fail, pause the exam and let the proctor know what has happened. They will be able to assist you in restarting the software. Then you can resume the exam with minimal loss of time.

The results of the exam

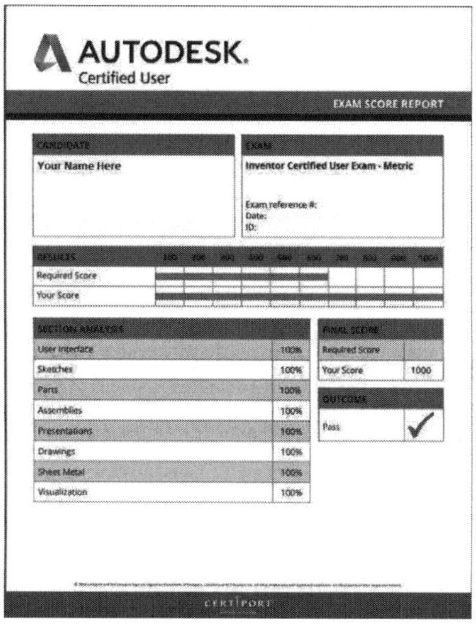

After completing the questions or when you reach the time limit you'll submit the exam for review. The review process is automated and only takes seconds. Once the score is calculated, you'll be given the result based on a scale of 1,000 being a perfect score.

You will also be able to review the details for each of the certification domains to see in a general sense how you performed on each part of the exam. If you pass the exam, this is good information to help you look at additional training or practice. If you don't pass the exam the first time, it is very valuable information to help you prepare for a retake.

If you want to review the results in the future, you can simply log into your account and select the exam to find the results information.

Retaking the exam

If you need to retake the exam you can begin a second attempt after waiting 24 hours. You will need to repurchase the exam each time you attempt a retake.

If you require a second retake attempt, you'll need to wait 120 hours. These delays are built in for the purpose of encouraging you to study the areas where your knowledge needs improvement. Any further attempts at a retake will have a five day waiting period between each attempt.

Additional resources

Exams and practice exams through Certiport: www.certiport.com/autodesk

Autodesk certification information: www.autodesk.com/certification

Instructor led training at Autodesk Authorized Training centers: www.autodesk.com/atc

Notes:

Chapter 3: What is Autodesk Inventor?

In the autumn of 1999 Autodesk introduced a new 3D design and engineering software named Inventor. Prior to the release of Inventor, Autodesk offered a popular set of design tools named Mechanical Desktop that could do solid and surface modeling, assembly modeling, and 2D drawings. Unlike Mechanical Desktop which had tools that used AutoCAD as a foundation, Inventor had its own code base. In the years since its release, the capabilities, power, and speed of Inventor have continued to increase and expand. Now every day, hundreds of thousands of people around the world create their products with this incredible tool.

The Inventor Workflow

At its core, Autodesk Inventor is an assembly modeling software. It can build large scale, fully articulated mechanisms made up of solid and surface models. It can create parametric solid, surface, and freeform models, but some of its greatest productivity tools exist in the realm of the assembly. Once the Assembly or components have been created, you can develop complete detail drawings of the design with Bill of Material information created directly from the assembly definition.

The first exposure most people have to Autodesk Inventor is in the creation of feature based solid models. The vast majority of people are creating solid models using a workflow called parametric solid modeling. Parametric solid modeling leverages sketches to form features that are also based on dimensional values, or parameters, whose size can be edited. At any time, the sketch can be edited to change the profile used for creating the feature.

This connection between precision of dimensions in the sketch and the role of those dimensions in machine components and products is logical for the user and allows for common changes in those components to be easily added to the parameters of the model. Relationships between features and components can also be represented in the parametric sketch and in the order of how features are defined.

Getting started

While people commonly begin creating Inventor models using a part model, Inventor works very well beginning with an assembly and in the context of that assembly. Once a part file is created, the vast majority of models begin with the sketch. The sketch must be placed on a plane either defined by the intersection of the X, Y, and Z axes, or on a plane that's been constructed in the model. The sketch can be formed in Autodesk Inventor with or without dimensions initially. Dimensions are used to size the sketch geometry, but it's the geometry that defines the shapes in the 2-D sketch used to create the 3-D model. Dimensions can be applied at any time after this geometry has been created. It's possible to create a 3-D model feature with no dimensions or constraints applied to the sketch. However, to be able to control the shape of this sketch, dimensions and constraints must be used.

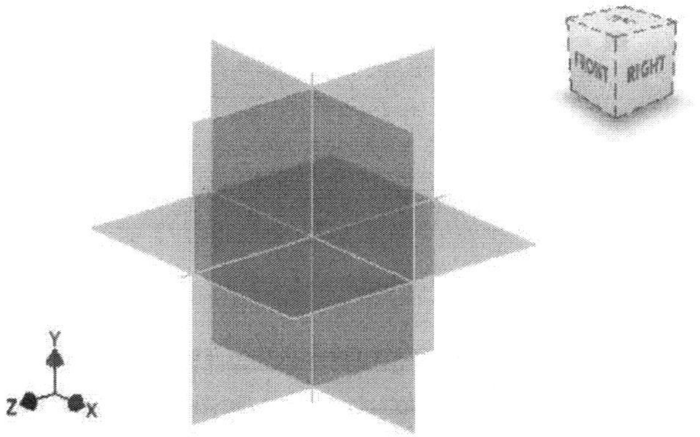

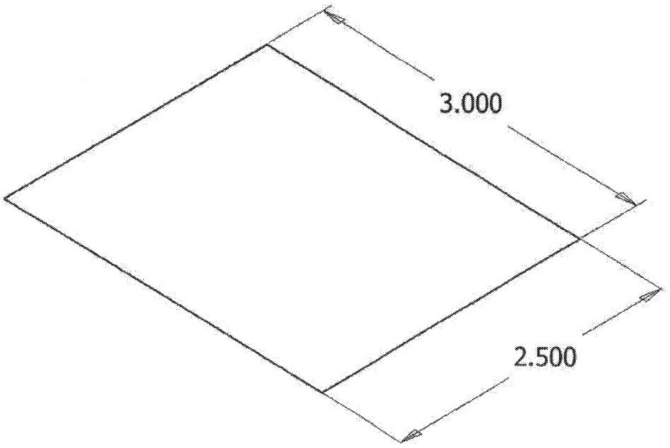

Part Models

A part model is a collection of features. This collection of features can represent simple real-world components or components of incredible complexity. Part models can be constructed in multiple ways. The most common of these and the focus of our attention in preparing for the user certification will be the parametric solid model.

For an example, think about a simple building block. That block has a length, a width and height. If you sketched the rectangle or square to define the width and length, the 3-D feature would then define the height. If you round the corners of the block using a fillet, that is captured by Inventor as a separate feature. Now, to add a hole to the center of the component you can position the hole feature based on the length and width parameters so that if the overall size of the block changes the hole can remain centered. Fillet features and most hole features do not use sketches. They are placed on existing models by selecting geometry to select where they will be placed.

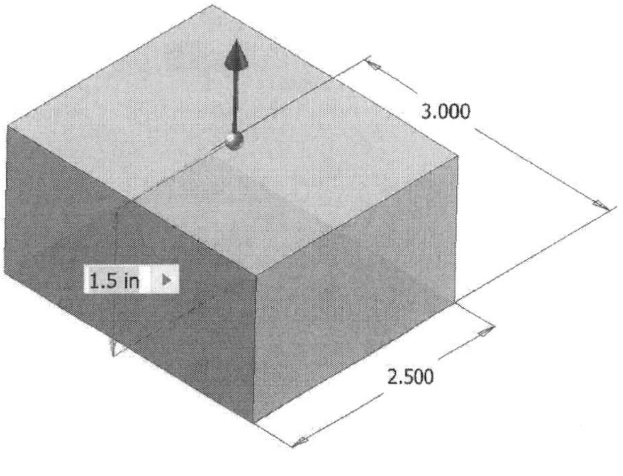

For every feature you create, Inventor will capture the size information and relationship to other features in the model. As you create these features, the history is captured in the interface and each individual feature can be accessed directly for editing with a double click on the feature in the browser or clicking on the feature in the design window and selecting edit feature from the mini toolbar.

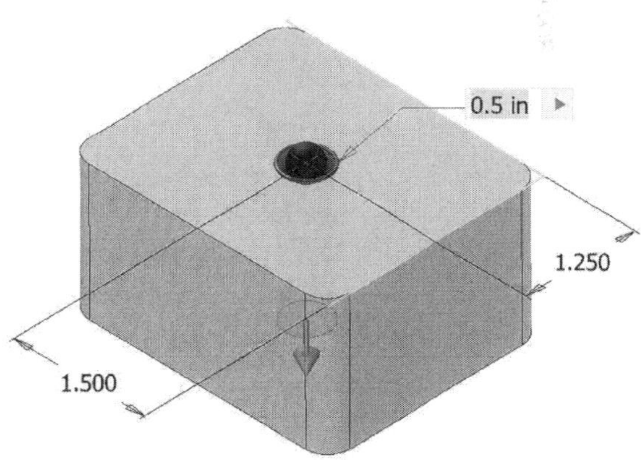

Editing a feature will bring up the same dialog box that was used to create it. This is a great feature for keeping the editing process simple and familiar. After editing a feature, the model will be regenerated. Some changes can cause other features to fail. You can repair the feature for the sketch that is failed by editing it individually. Inventor can continue to model with many failures. Sometimes this is necessary for significant changes to the model.

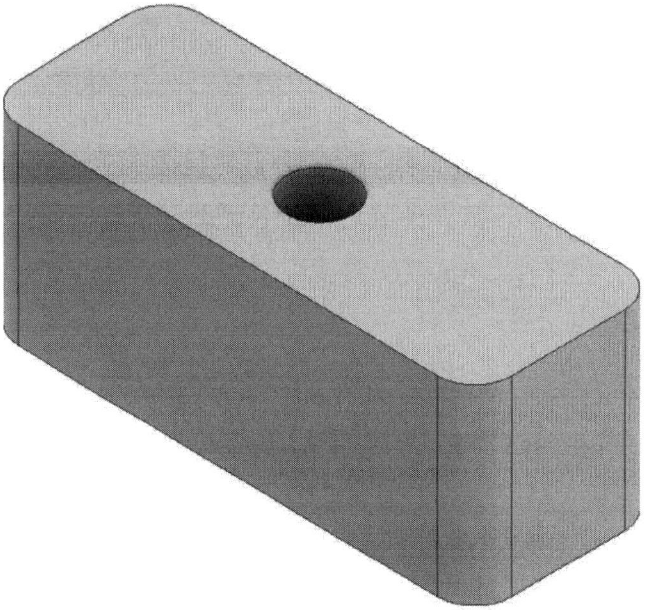

Assembly Models

Assembly models are collections of parts and other assemblies held in position through the use of constraints. Flexibility can be built into the assembly to simulate real world conditions or parts to make it easier to explore the range of motion in mechanisms.

Every part or other assembly that is in the assembly is linked to an external file whether the part or sub-assembly was created originally in the context of the active assembly file. This approach to external files is the norm for Autodesk Inventor and its competitors. This approach allows the design system to offload feature information

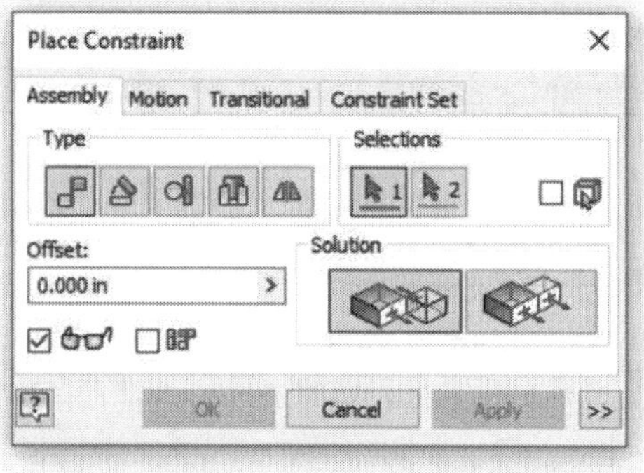

when working in the assembly and load the feature information back into memory when actively editing a part in the context of the assembly.

There are a wide variety of constraints that offer varying levels of control between faces, edges, and points in the model. Each constraint carries with it a level effect on the degrees of freedom of components in the assembly. Every component placed in an assembly has 6 degrees of freedom - three translational and three rotational. By removing degrees of freedom, constraints are used to replicate how parts are stacked, bolted, welded, or held together in some other fashion.

The potential motion of the constraints can be limited either by setting a limit on the range of motion or detecting when components come in contact with one another. Constraints can also include offsets for creating clearance between components. These offsets can also be represented with negative values.
Another feature of the assembly environment is the Content Center. In the Content Center, you can find millions of standard components like nuts, bolts, bearings, and standard steel shapes. These objects are available for selection and use in standard, purchasable sizes.

Detail Drawings

For many users of 3-D CAD, the ultimate goal is to produce a 2-D drawing. These drawings can by created from the individual part files or from the assembly file.

Drawings can contain orthographic projections showing the front, right, left, top and bottom of the model. They can also include isometric views, section views, detail views, and auxiliary views.

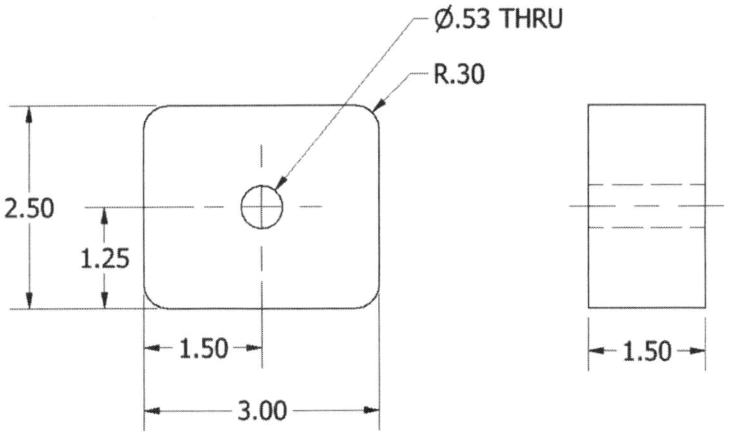

The typical workflow is to place a base view on the drawing that best represents as much information on the design as possible. Then, views are projected from that base view. Any view can be used as a parent view creating any number of child views. If a parent view moves orthographic projections of the view, it will also move to maintain their proper position. You can move a projected view but by default it will maintain its position to its parent view. Drawing views need to be scaled to fit within the border of the sheet. The scale can be changed to accommodate the number of views required to properly display the critical information of the design. Changing the scale of a parent view will update the scale of views projected from it.

When creating views of the single part, hidden lines are automatically generated in the drawing view to represent features that are obscured in the viewing position. By default, drawing views of an assembly did not display hidden lines. The display of hidden lines or showing a drawing view that contains color shading of the part can be changed at any time.

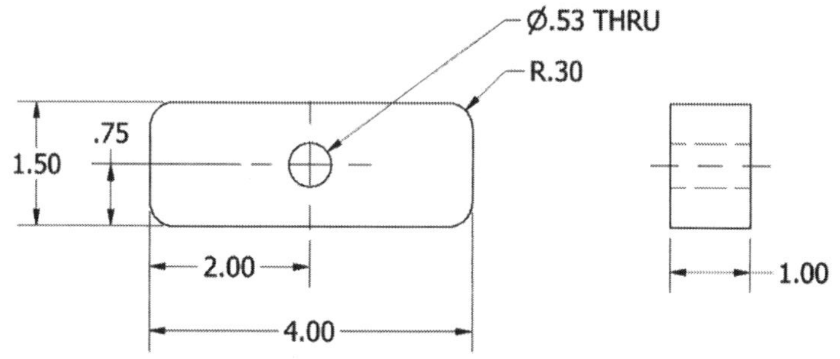

After drawing views have been located on the sheet, detail dimensions can be applied. These dimensions represent the sizes of the 3-D geometry regardless of the scale that the view was created in. A single dimension tool is capable of applying vertical, horizontal, aligned, radial, and diameter dimensions without having to restart the dimension tool. All of these dimensions will be associative to the geometry of the model so that if the model is changed, the dimensions in the drawing will update automatically.

There are specialized detailing tools for calling out hole notes, surface finish, and geometric dimensioning and tolerancing symbols. For the assembly, balloons can be added that will have their numbers automatically populated based on an item number in the bill of materials. The bill of materials information is also extracted to populate the information of the parts list.

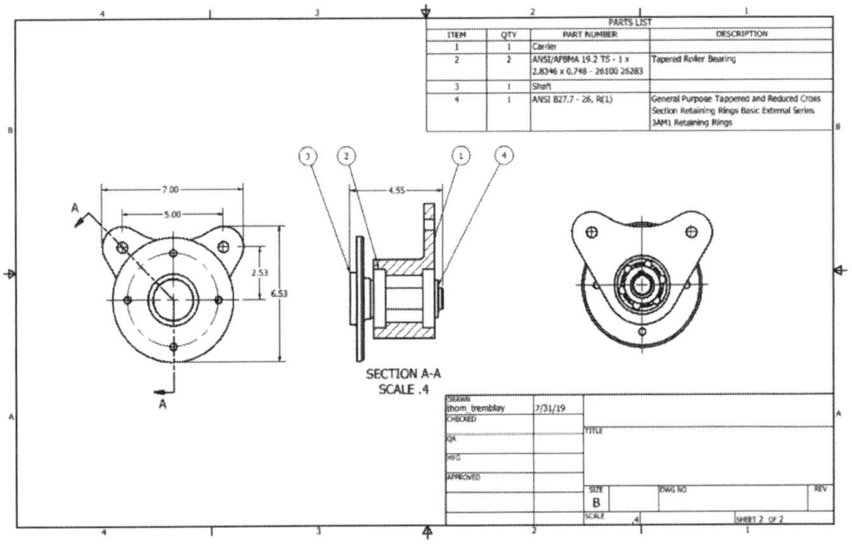

Current Autodesk Inventor System Requirements

Autodesk Inventor is an exceptionally powerful design tool. To fully leverage its power requires the proper hardware and operating system. With this said, Autodesk has gone to great lengths to make it possible to get very good performance from Inventor on affordable computers. Below are the specifications as listed on the day of this writing from the Autodesk website.

Operating System	64-bit Microsoft® Windows® 10. See Autodesk's Product Support Lifecycle for support information.
CPU	**Recommended:** 3.0 GHz or greater, 4 or more cores **Minimum:** 2.5 GHz or greater
Memory	**Recommended:** 32 GB RAM or more **Minimum:** 16 GB RAM for less than 500-part assemblies
Disk Space	Installer plus full installation: 40 GB
Graphics	**Recommended:** 4 GB GPU with 106 GB/S Bandwidth and DirectX 11 compliant **Minimum:** 1 GB GPU with 29 GB/S Bandwidth and DirectX 11 compliant See the Certified Hardware

Display Resolution	**Recommended:** 3840 x 2160 (4K); Preferred scaling: 100%, 125%, 150% or 200% **Minimum:** 1280 x 1024
Pointing Device	MS-Mouse compliant Productivity: 3DConnexion SpaceMouse®, driver version 10.7.0 or later.
Network	Internet connection for web install with Autodesk® Desktop App, Autodesk® collaboration functionality, web downloads, and licensing. Network license manager supports Windows Server® 2016, Windows Server 2019 and the Windows 10 desktop versions listed above.
Spreadsheet	Full local install of Microsoft® Excel 2016 or later for workflows that create and edit spreadsheets. Inventor workflows that read or export spreadsheet data do not require Microsoft® Excel. See Inventor Excel Requirements for more information. Office 365 subscribers must ensure they have a local installation of Microsoft Excel. Windows Excel Starter®, OpenOffice®, and browser-based Office 365 applications are not supported.
Browser	Google Chrome™ or equivalent
.NET Framework	.NET Framework Version 4.8 or later. Windows Updates enabled for installation.

For Complex Models, Complex Mold Assemblies, and Large Assemblies (typically more than 1,000 parts)	
CPU Type	**Recommended:** 3.30 GHz or greater, 4 or more cores
Memory	**Recommended:** 64 GB RAM or greater
Graphics	**Recommended:** 4 GB GPU with 106 GB/S Bandwidth and DirectX 11 compliant See the Certified Hardware

Trial software access

To prepare for the Autodesk Inventor Certified User exam, it's important to spend time in the software. If you don't already own the software, Autodesk Inventor can be purchased or downloaded for a free 30-day trial at www.autodesk.com/inventor.

Students and teachers can gain free access to Inventor and many other software packages by registering at the Autodesk education community which can be found on the web at www.autodesk.com/education.

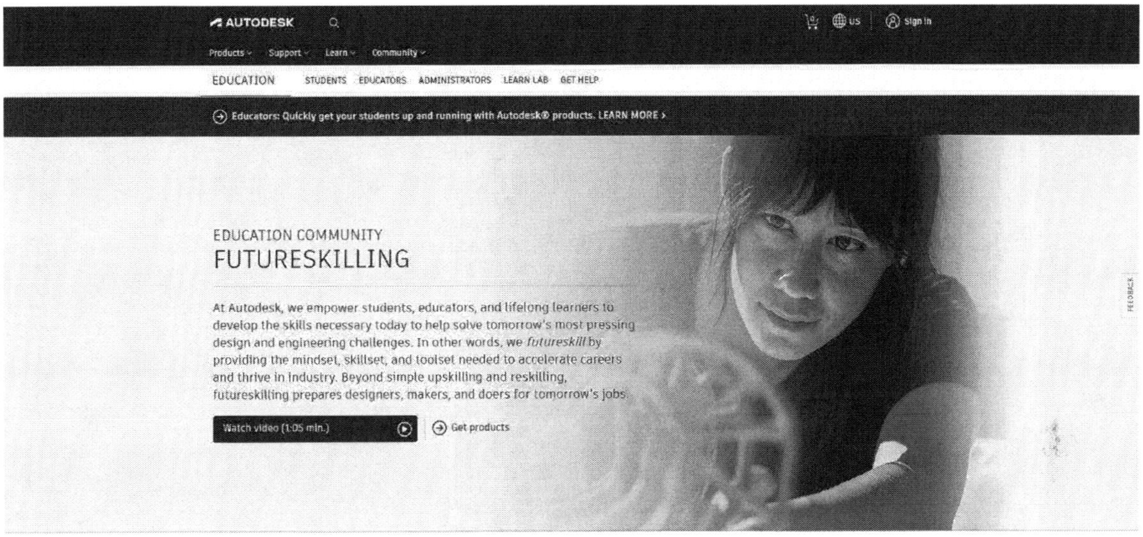

Notes:

Chapter 4:
User Interface and Navigation objectives

The user interface is your connection to Autodesk Inventor and its design, engineering, and manufacturing capabilities. The user interface will change as you edit different types of files, and the hundreds of tools in Inventor are organized in the toolbar in tabs and panels.

Tabs can be looked at as collections of tools that serve a generally similar purpose. Examples of tabs are 3D Model, which is for creating and modifying the features of a 3D solid model, and Assemble—this is where you find the tools for joining and constraining components together to build an assembly. A tab can also be temporary. The Sketch Tab only appears when you are editing a parametric sketch.

Tabs

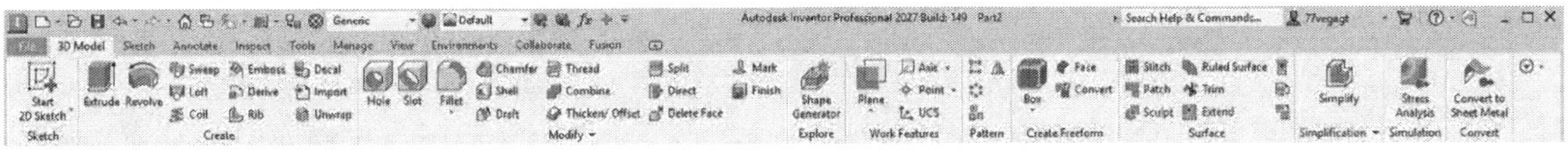

Panel

A panel is a subgroup of tools that are more closely related to one another. Within the 3D Model tab, separate panels divide solid model creation, editing, and patterning tools.

Toolbar tabs and panels can have their visibility switched off and on to refine the interface to suit your specific needs.

Change the viewpoint using the ViewCube

Navigating the Autodesk Inventor interface is made easier thanks to tools like the ViewCube.

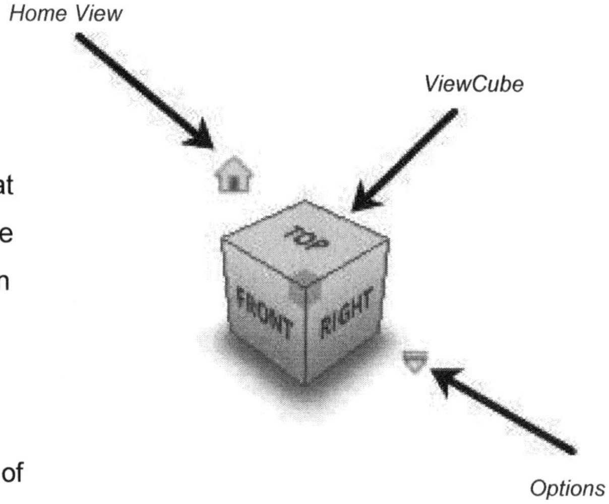

The ViewCube not only allows you to use the mouse to rotate your model, it also establishes the orientation of your model for creating drawings. The view of the model that presents itself when looking at the front of the ViewCube will be the Front view generated in the drawing.

Selecting the named faces, edges, and corners of the ViewCube will rotate the view of the model in the design window. When the model is positioned so that you are looking directly at a face, additional controls will appear to pivot the model around the axis of the current view.

You can always restore the model to the original view by clicking the Home View icon near the ViewCube. Selection can be made with a mouse but when you're working with a touchpad or mouse without a wheel the ViewCube can be especially useful.

Change settings of the ViewCube

If you select a face on the ViewCube to establish a model view position, you can set that view to be oriented as either the top or front view on the ViewCube. This is done by setting the view position and using the mouse to right-click on the ViewCube and choosing to make that view either the Top or Front depending on your preference.

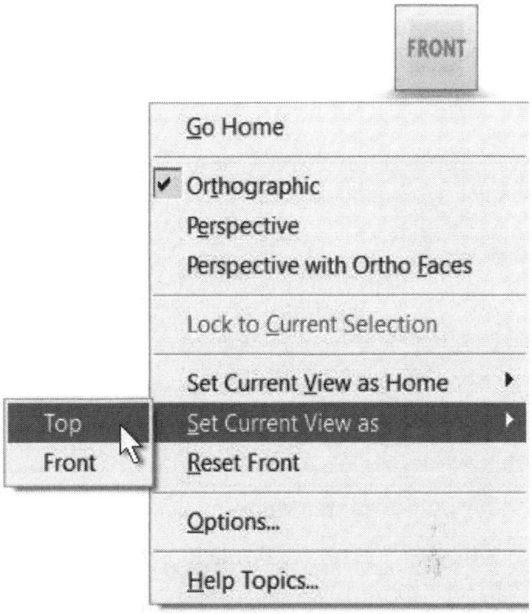

Another option in the menu is to set the current view orientation of the model to be the Home view so it can be quickly recalled using the Home View icon near the ViewCube.

The context menu for the ViewCube can also set the view mode of the model to be Orthographic, Perspective or Perspective with Ortho Faces. This last option will display the model in perspective unless you choose a face on the ViewCube. Selecting a face will set the display to Orthographic once the view is set.

Orthographic view mode

Perspective view mode

Understand Inventor file types and standard templates

A complete design in Autodesk Inventor can use many different files and different types of file to construct the design. Each type of file uses a different template and is stored with a different file format.

Part file = .IPT

Assembly file = .IAM

Presentation file = .IPN

Drawing file = .IDW

Drawing file = .DWG

An assembly can use individual part files which use a file extension of .IPT. Assemblies can also use other assemblies which are saved with an .IAM file extension. Presentation files are commonly used to create exploded view animations of the assembly file using the .IPN file extension. 2D detailed drawings can use either the .DWG or .IDW file type and can generate views of Part, Assembly, or Presentation files. Each of these formats displays a different icon on the computer to make locating the type of file you're looking for easier.

Autodesk Inventor can also import and export many different file types either in neutral formats like .STP or .IGES, or CAD software specific formats created by other leading design software. These files can be imported directly as models without a feature tree, or you can have Inventor search for and rebuild common parametric features from the model.

There are other approaches for connecting data from other systems into Inventor while allowing them to be maintained in their native formats.

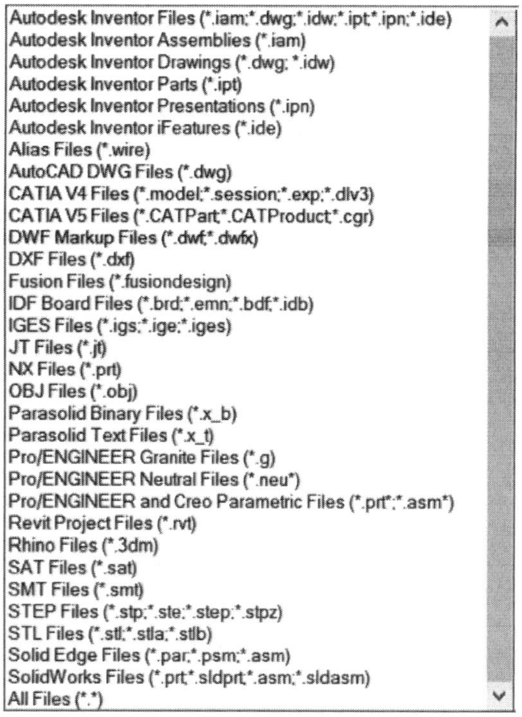

Chapter 5: Sketching Objectives

Because of the parametric basis of much of Inventor's solid modeling capabilities, sketching is one of the most important aspects to fully understand and can be more challenging than people might initially suspect.

The Sketching objectives for Inventor user certification cover a broad variety of the tools to manage and set the size of geometry in the sketch.

Apply dimensions to a sketch

Manufactured products must be made to specific sizes to ensure fit. Inventor uses parametric dimensions to set the size of sketch geometry that the base features of the 3D model is built from.

This approach separates the function of the shape of the geometry which is still handled by sketch object tools from the size of that geometry which is controlled by dimensions and constraints. After roughing in the sketch, dimensions are placed on the geometry to set its size. These dimensions can work independently or relate to one another to perform functions within the sketch.

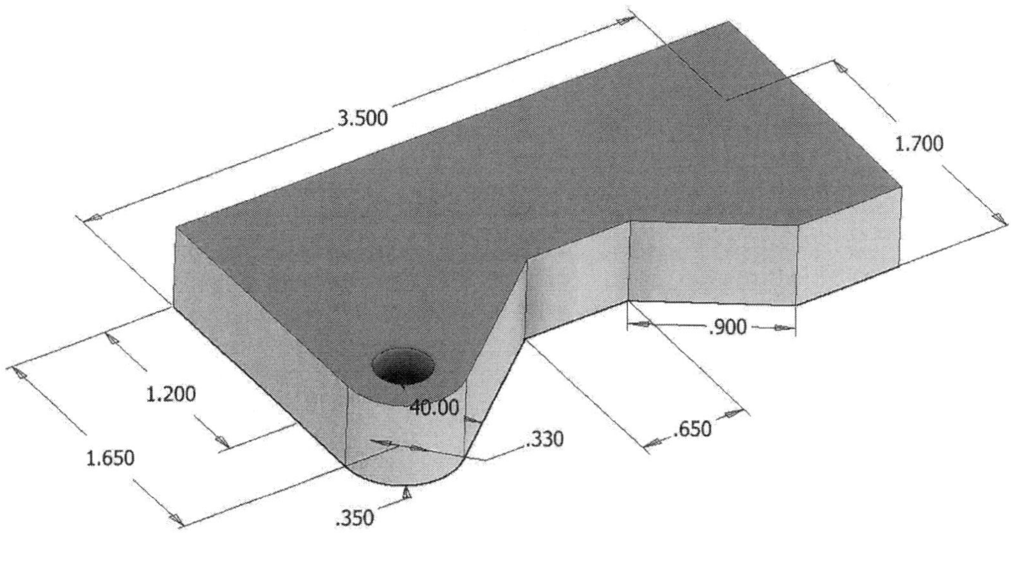

Inventor only requires one tool for adding dimensions to a sketch. After starting the tool, selecting a single edge will apply the dimension to that edge, but you can also select additional points, arcs, circles or edges to dimension the geometry as needed. Depending on the sketch geometry selected, the dimensions that will be placed can be linear, aligned, angular, radius or diameter.

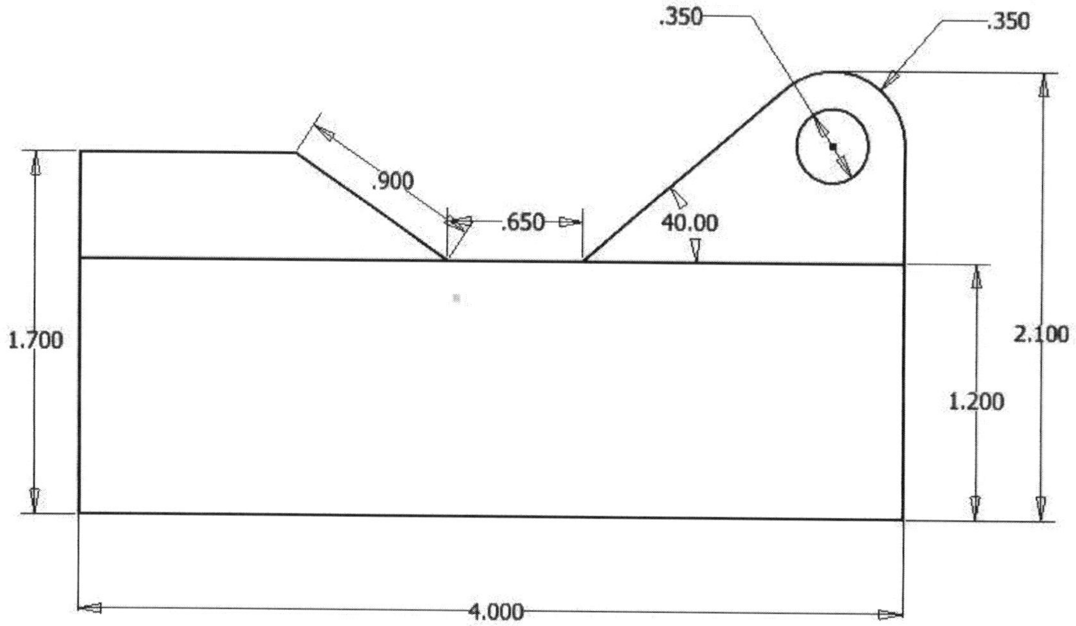

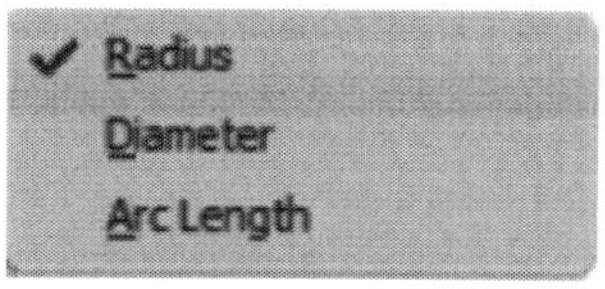

Using right-click context menus, you can change the initial type of some dimensions. For example, after picking two points it might offer a horizontal or vertical dimension, but with a right-click you can tell the system to apply an aligned dimension. After picking to place an arc, Inventor will offer a Radius dimension, but using the right-click menu you can choose to place a Diameter or Arc Length value instead.

Assign geometric constraints

In addition to dimensions, the other primary way of controlling a sketch is through geometry relationships referred to as constraints. These relationships are not as easily modified as dimensional relationships and therefore are more enduring.

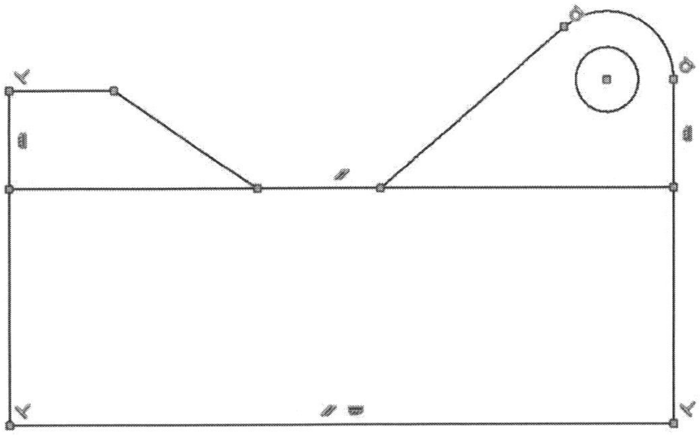

Each constraint has specific properties and uses. Some constraints can be used as an option to others depending on the use of the sketch and how you anticipate future changes.

Coincident - Points can have their position associated with other points, lines, or arcs. This serves to maintain position relationships and to close sketch loops to allow for creation of 3D geometry.

Collinear - Keeps line segments aligned in the same direction.

Concentric - Associates the center points of arcs and circles by selecting two curved edges per association.

Fixed - Any object in the sketch and any number of objects can be fixed in place.

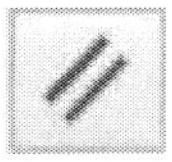

Parallel - Line segments will maintain their parallel relationships regardless of the angle, length, or distance apart.

Perpendicular - Line segments will maintain their perpendicularity regardless of the angle, length, or distance apart. These segments do not have to be in direct contact with one another.

Horizontal - May be applied to edges and will hold them in a position relative to a sketch axis. This constraint can also be used to position points relative to other points or edges while simultaneously related to the "horizontal" axis. When applying this constraint, a dotted line will indicate the direction that is perceived as horizontal in the sketch plane.

Vertical - May be applied to edges and will hold them in a position relative to a sketch axis. This constraint can also be used to position points relative to other points or edges while simultaneously related to the "vertical" axis. When applying this constraint, a dotted line will indicate the direction that is perceived as vertical in the sketch plane.

Tangent - Creates a tangency condition between an arc or circle and other geometry. These objects do not have to be in contact with each other to remain tangent.

Smooth - Used to build continuity between curved segments. This can include straight segments. Similar to Tangent but builds a continuous curvature (G2) relationship that takes the curvature of adjacent segments into account for calculating the transition.

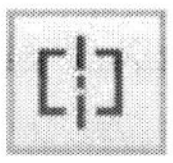

Symmetric - Connects sketch elements on either side of a Symmetry Line. This constraint will affect the position of one object relative to another but not necessarily control the size.

Equal - Making objects equal in size can replace many common dimensions in a sketch used to create consistent sizes.

Project geometry

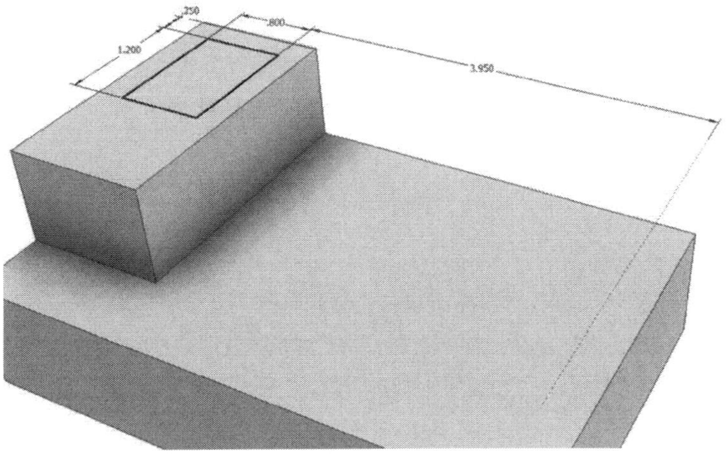

When creating a sketch there are times where existing geometry in the model can be useful for reference. Edges, points, faces or construction geometry in the current model can be projected into the assembly to offer that reference or to create the sketch elements to be used for the 3D feature itself.

You can project these edges onto the active sketch or even project the edges of a body where it intersects the active sketch plane. Once geometry from the model is projected into the plane, any changes to that geometry are kept up to date in the sketch to make sure the reference is still relevant.

An option for any type of geometry in the sketch that can be very effective for projected geometry is to make that geometry construction geometry. The Construction toggle is in the Format panel of the Sketch tab. Construction lines will be dashed and a different color. The only difference between construction and normal geometry is that construction geometry is ignored for defining the profile of a sketched feature.

Create and modify geometric shapes

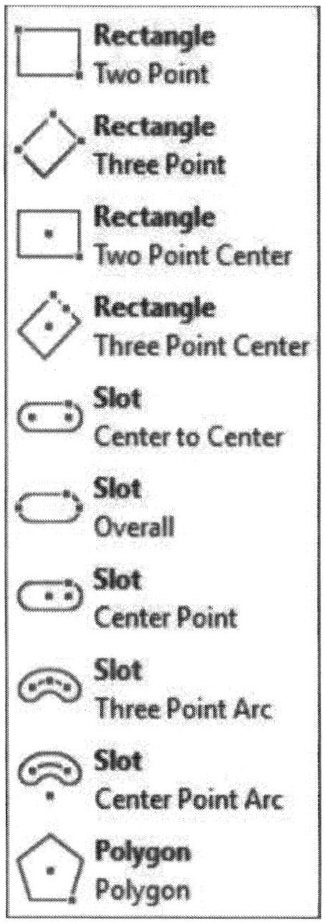

The process of creating a sketch in Autodesk Inventor is generally like creating a sketch in any CAD software. Like other software, these sketches can be made up of lines, arcs, splines, and circles. There are also shortcuts to creating specific types of rectangles, or special geometry like slots or polygons.

These sketches will also include geometric constraints that are added to the sketch in any intelligent parametric dimensions. Another difference is that sketches in Autodesk Inventor can have multiple profiles in the same sketch which can each be selected to create 3D geometry.

Once the sketch is defined, you're always able to go back and make modifications to the geometric constraints or change the values of the dimensions to modify the size and shape of geometry in the sketch. If you've built 3D features from the sketch, those features will update to reflect the changes once the editing is complete.

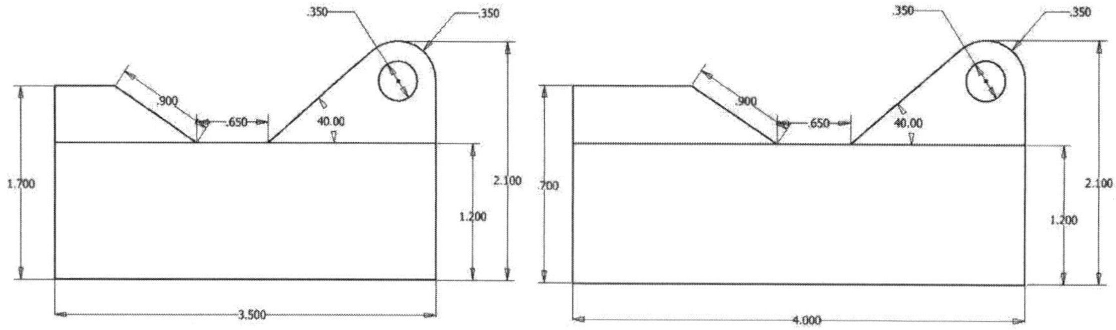

Modify an Inventor model

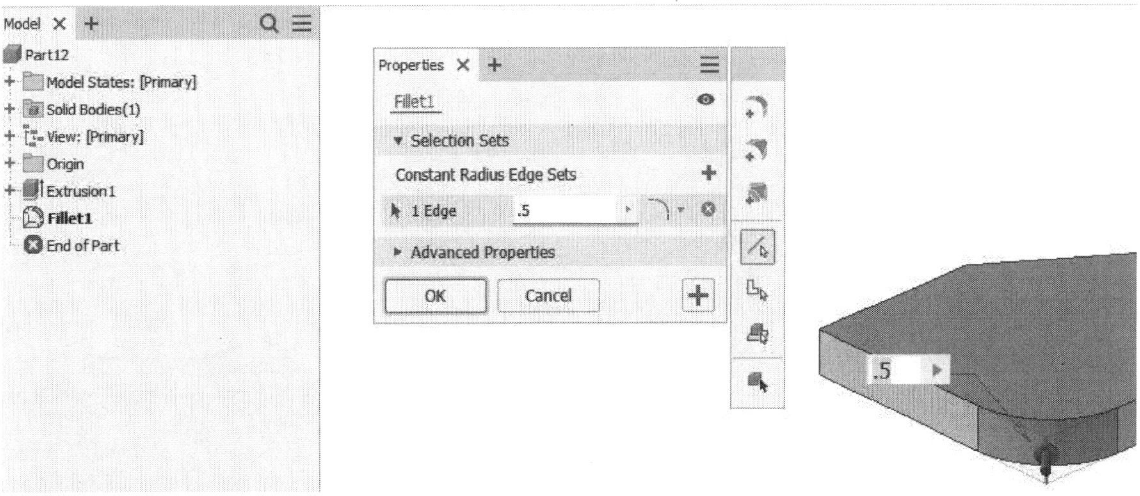

Editing the features of the model is as easy as double-clicking a feature in the browser, using a right-click on the feature in the browser (the text of the edited feature will become bold) then selecting Edit, or selecting a feature in the design window and choosing edit feature from the mini-toolbar that appears.

For sketched based features you'll also have an option to edit the sketch from the context menu. This will return you to the sketch to make modifications to the dimensions or constraints. For placed features it will reopen the dialog showing the same preview that was presented when the feature was created so you can review the properties of the feature and make modifications.

If the feature is not the last in the model, the features that came after it will be suppressed while the edit is being made. After changing any sizes, termination options or other settings, the model will update after clicking OK to complete the modification.

Notes:

Chapter 6: Part Modeling Objectives

Working with solid models effectively requires proficient use of the tools called out in the Certification objectives. These features build the initial solid shapes in a part (.ipt) file, add details and quickly replicate common features.

These features have many options and can be defined in different directions and offering very different results.

Create extrude features

Features based on sketched or projected profiles are categorized as sketched features.
A commonly used sketched feature in solid modeling is the extrude feature. This type of feature adds height to a sketched profile.

An Extrusion can be created above or below the plane of the sketch in a single direction, symmetrically in both directions at once, and in both directions using different methods including the ability to continue building the feature until it meets another feature of the model.

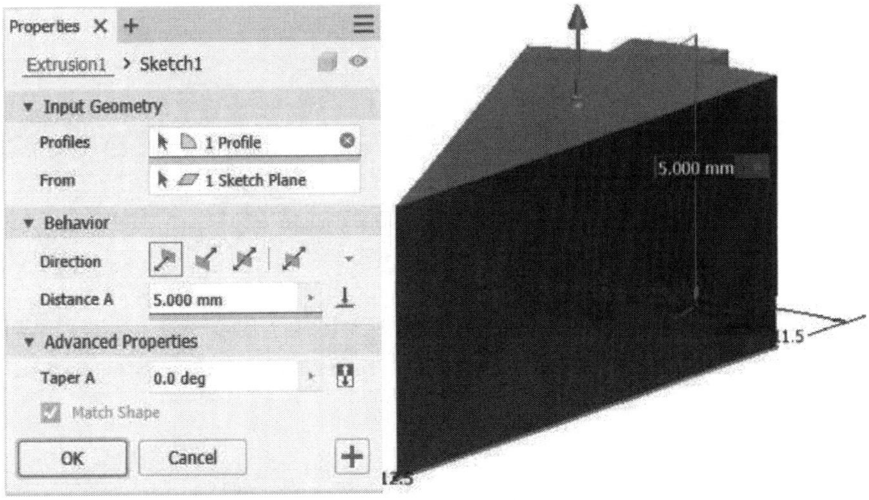

Once a height is entered for this feature, that value is a parameter in the model and can therefore be linked to other model parameters. An option when extruding is to add draft or a taper angle to change the size of the profile over the distance the extrusion is created. This is useful for castings and plastic components.

Extruded features can add geometry to the model, remove geometry from the model, or form 3-D geometry by intersecting the shape of the sketch with existing geometry in the model.

Any extruded feature will update automatically if the sketch that defined its shape is changed. A sketch can be shared to create multiple extrusions from it and any number of extrusions can be created in a file.

Create a pattern of features

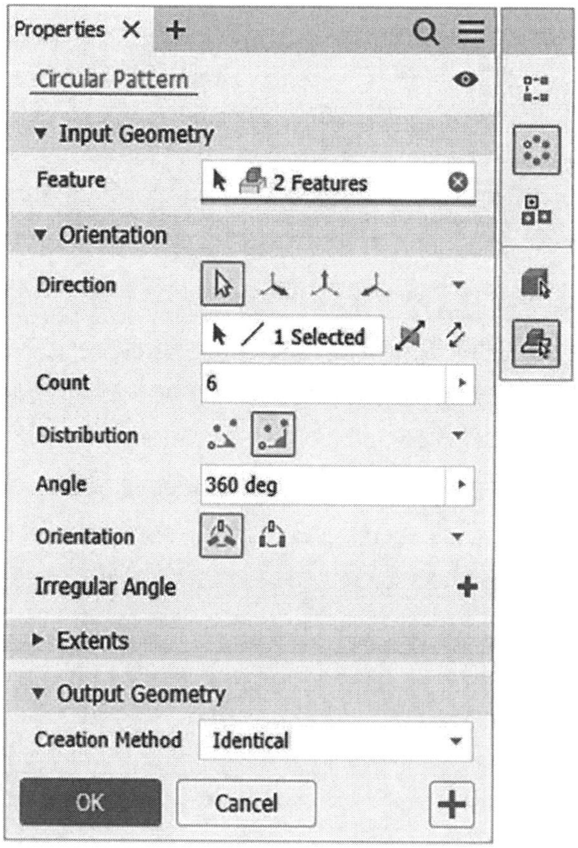

Many manufactured components have shapes that are repeated in circles, columns and rows or along a path. In Inventor these features are called patterns and they can be created in several ways.

A Circular pattern replicates one or more features any number of times around an axis. To create the pattern, select the features you want to duplicate, then switch Inventor to select the axis by clicking the arrow icon next to axis. Then enter the number of patterned features before clicking OK. Depending on the feature options you can create this pattern in a complete circle spaced evenly over an angle or spaced apart at specific angles. A pattern can also be created in either or both directions around an axis. The features that are patterned can maintain their orientation to the sketch or reorient radially to the axis as they are created.

A Rectangular pattern can create a pattern in up to two linear directions at the same time and though the name of the tool is "rectangular", these two directions do not have to be perpendicular. The instances of the features can be duplicated based on a space between each of them or spaced equally over a distance. An additional option is to create the pattern along a path. This path could be a series of line segments or a curve. The extents and the spacing of the pattern work like a Rectangular pattern with the option of using the curve length as a limit on the distance the pattern will go.

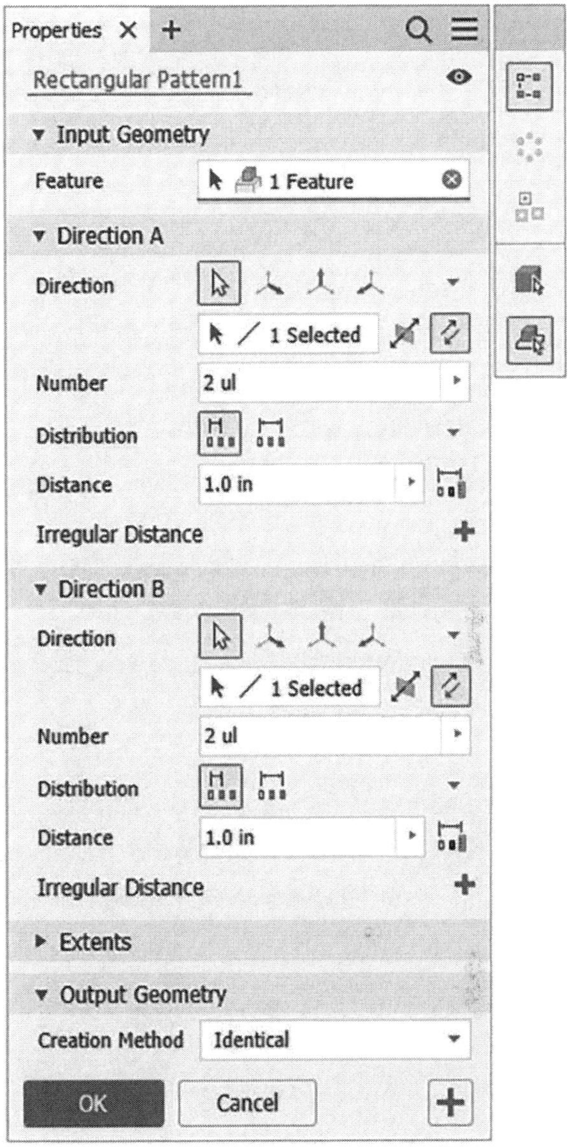

The last type of feature pattern is the Sketch Based pattern. A feature or features can be quickly duplicated using points in a sketch for placement. There is an option for specifying a base point on the feature to align to the sketch points. If the sketch is later edited, the features will be repositioned to maintain their connection to the points in the sketch.

Create a shell feature

If you need to make a solid model hollow, an easy way to do it is to apply a Shell feature. The primary workflow is to select a face or faces that will be removed from the solid and then to set a consistent thickness value for the remaining faces of the model. You will then need to choose to add the thickness of the remaining faces to the inside, outside, or evenly divided on both sides of the original face.

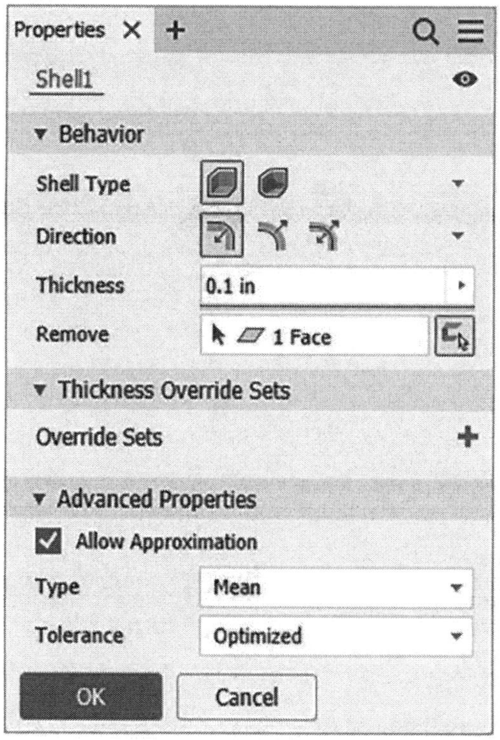

Expanding the Shell Feature dialog will give you the option of selecting individual faces to apply a different thickness. Multiple faces can be selected for additional thickness values.

A Shell feature can also be created without selecting a face to create a hollow body with no opening. Like other shell features, the thickness can be added to either side of the exterior body or have the new thickness divided across the original face.

Apply fillets and chamfers

Fillets add a rounded face to a 3D model. Most commonly they're added to what were originally sharp edges. There are many options for adding fillets to the model.

In one fillet feature you can select as many edges as you want and apply a single radius value to them whether those edges are on outside or inside edges of the model. Options in the dialog allow for choosing loops of edges, and features can be used as well as selecting edges. There is also an option for selecting all of the inside (fillets) or outside (rounds) edges at once. You can also apply multiple radius values to different edges of the same model or feature.

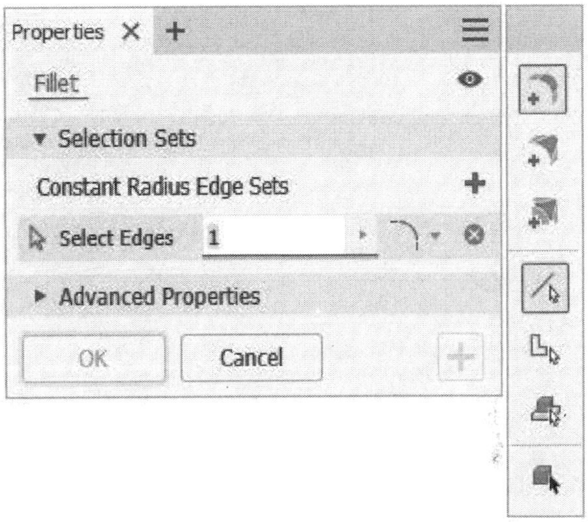

These fillet features themselves have three different ways of being calculated. Tangent fillets are developed with a consistent radius tangent to the adjacent faces. G2 Fillets create a continuous curvature taking in the orientation and size of the adjacent faces.

Edge fillets do not have to have a consistent radius. You can also create fillets with variable radii by establishing a beginning and end radius for the selected edge. After selecting the edge, you have the option of picking points along the edge and setting another radius to transition through. Each one of these points has a radius value and a position value which is based on a portion of the length of the edge expressed by a total of 1. A transition point placed half the length of the edge is at .5 of the edge.

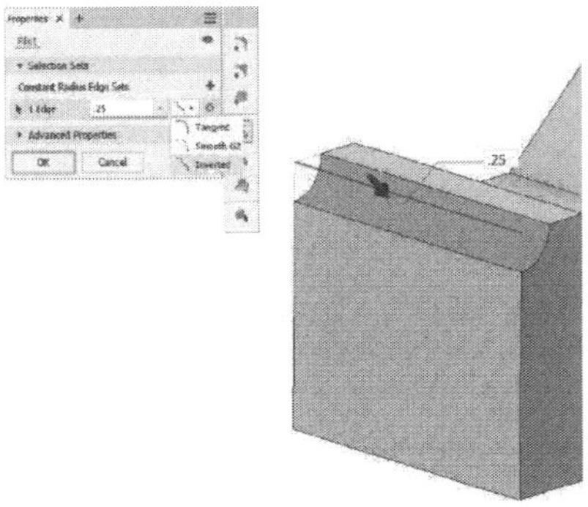

Edge fillets also do not have to form a rounded edge. The third option for edge-based fillets is the Inverted fillet which creates a fillet with the edges normal to the adjacent face.

When edge fillets converge in an external corner, the shape is a single face based on the geometries of the fillets. You can choose to add a setback by selecting the vertex of the intersection of the edges and setting a distance down each edge where the transition begins to form the corner.

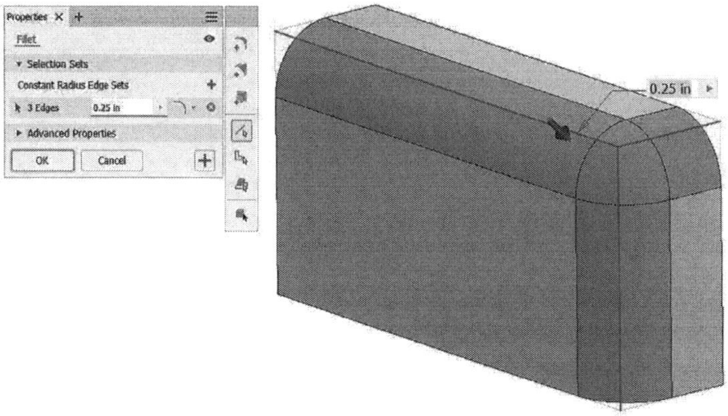

Inventor has two additional ways to create fillets: the Face fillet and the Full Round fillet. The Face fillet is used to converge two faces based on a target radius. These faces do not need to touch or have a consistent edge. This fillet is useful for filling gaps or creating unique solutions based on a radius.

The Full Round fillet creates a solution without specifying a radius by generating a fillet tangent to three faces regardless of what the resulting radius or variable radius fillet is required to connect them. While it can be used to create incredibly complex solutions in 3D, it is also used to round out ends rather than calculating edge fillets to half a face width.

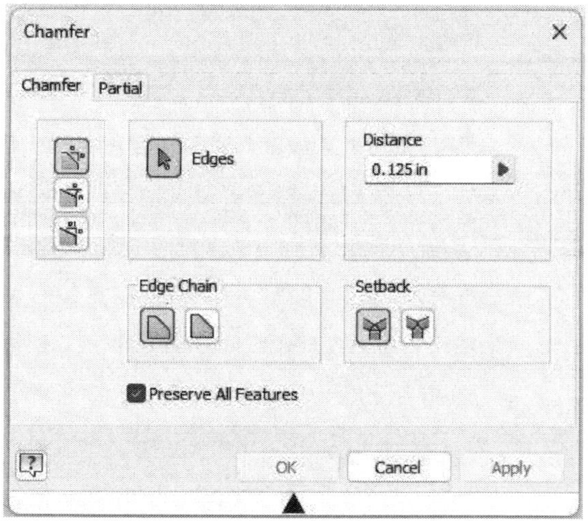

Chamfers apply a flat or beveled face to what was a sharp edge. These faces can set the original edge location back the same amount on either side of the selected edge, two different distances, or by setting one distance and then an angle for the new face from the original adjacent face. Normally the chamfer is created on the entire length of the edge, but an option is to place a partial chamfer holding the creation back from the ends of the selected edge based on an offset or the chamfer's length. Multiple edges can be selected for Chamfers, but all selected edges will get the same treatment in a single chamfer feature.

Create hole features

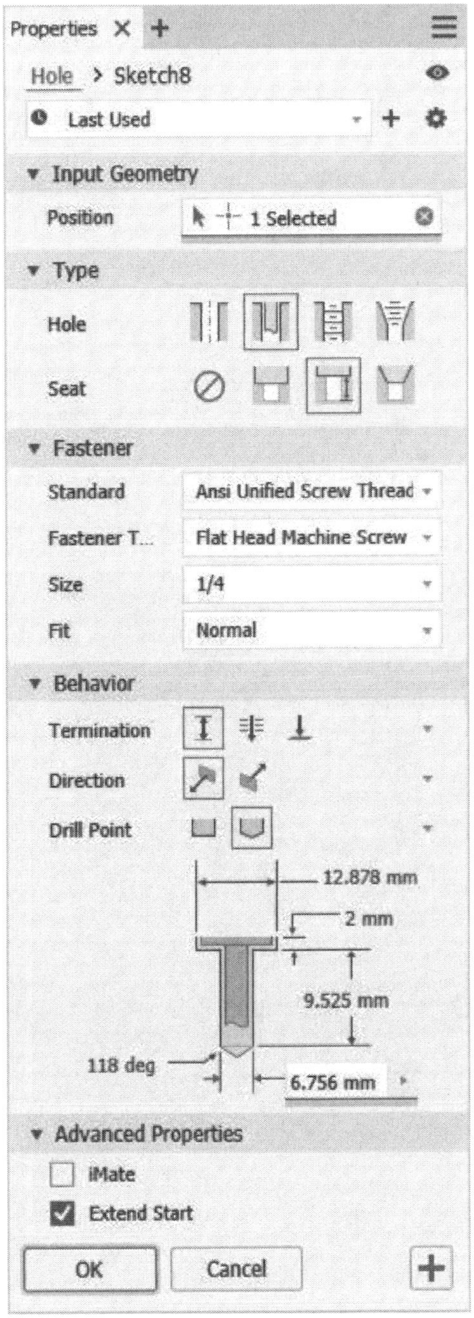

Some people might choose to create holes based on a sketched circle, but using the Hole feature offers many advantages of that approach. Holes can be placed on a face by picking a location, selecting one or two edges to define an offset from, picking a radiused edge or face to be centered on or picking a work point and then an edge or axis to set the angle for the hole. Another option is to place holes on the points of a sketch. This makes it easy to place many holes in one feature and is ideal for creating a pattern of holes placed on a sketch that is easily edited.

Choosing how to place the hole feature is just the beginning. Once the hole location is selected you will need to choose what type of hole it will be. A simple hole just uses a diameter and a termination type, while a clearance hole derives its size from a table of standards built into Inventor that places the correct sized hole depending on the size and type of fastener that will go through it. A Tapped hole has threads that are also based on standards and will present a list of available thread designations and classes after you've selected a nominal size from the pull-down menu in the dialog. The last option is the Taper Tapped hole which is most commonly associated with pipe threads or threads for self-sealing fittings. The threaded portions can be the full depth of the hole or based on a dimension from the beginning of the hole.

There are several Seat options for holes as well. The Simple hole starts its diameter from the surface selected to place the hole on. Counterbore and Spotface holes have cylindrical bores whose diameter and depth are created before the hole begins. The difference between the two is where the depth of the hole is measured from. A Counterbored hole's depth begins from the placement plane and a Spotface hole's depth begins at the bottom of the bore. The last seat option is the Countersink which uses a conical face to begin the hole. The angle of this surface can be changed.

Holes can be set to pass all the way through the component or stop when it reaches a selected face. They can also be set to a specified depth. An angle can be set for the bottom of the hole to represent drilled holes or holes created using a boring operation or for holes created by additive manufacturing.

At times holes are created in an orientation where model geometry overhangs or obscures their openings. The option to extend the start of the hole will take the shape and size of the hole start and reverse it back through the model.

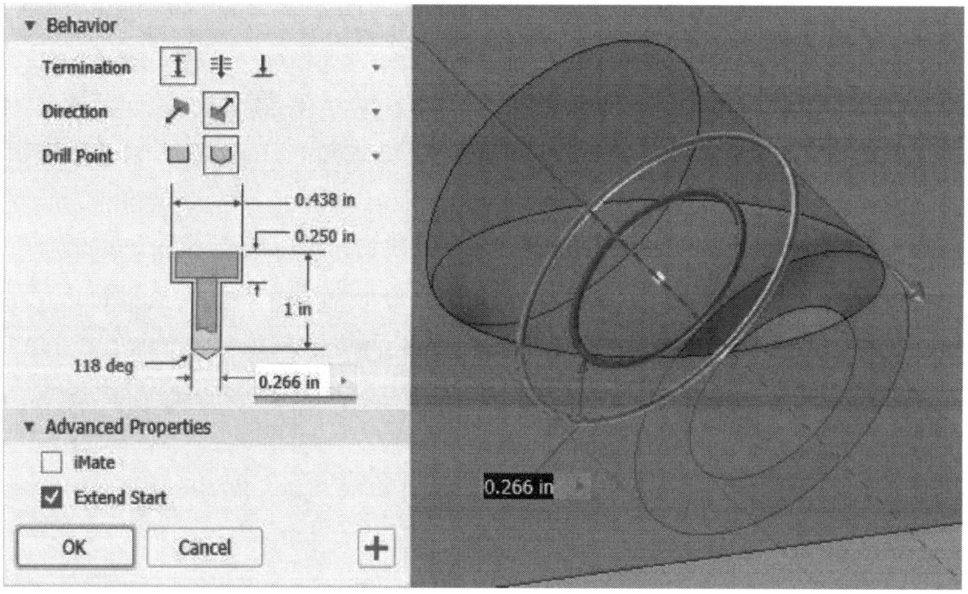

Create revolve features

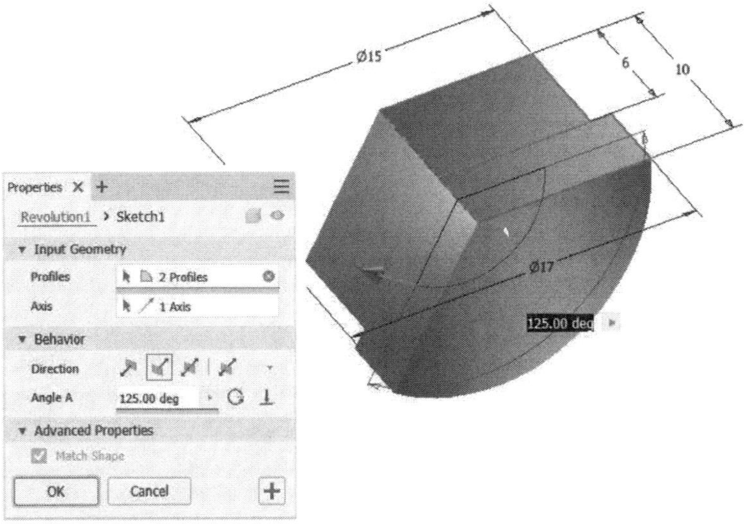

Objects with a consistent profile around an axis are ideal features for the Revolve tool. In machine design you'll frequently see this type of feature forming things like wheels or shafts.

Like an extruded feature, revolve features are based on a sketch but rather than building 3-D geometry in a linear direction, this geometry is created in 3D around a selected axis. This geometry can be created in one direction on either side of the sketch plane, both directions at the same angle, or formed on each side of the sketch plane using different angles or termination types.

These features can add or remove material from existing features, intersect with them or create a new solid in the design file. You can also choose to revolve the feature as a surface rather than a solid model.

Place threads

Hole features can have threads applied to them in the process of defining the hole. If you have a shaft or boss that needs to be threaded or even a cylindrical extrusion that was cut into the part, you can add a thread to that feature manually.

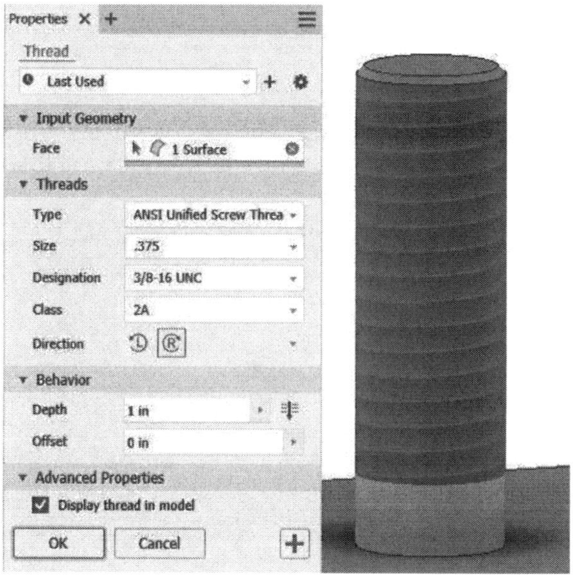

Once the cylindrical face is selected, choose the thread type, size, designation and other details. The thread can begin at the edge or be offset and can either be applied to the entire length of the face or a specific distance.

Thread features are presented on the model as an image that makes it clear the thread is applied but does not create a physical thread in the model. In a 2D Drawing, the major and minor diameters of the thread are displayed in drawing views and the proper information for a thread note is extracted and displayed in the note automatically.

Chapter 7: Browser Editing Objectives

The Browser is the control center for managing the content and structure of an Inventor design file. Whether it's the features of a part, the structure of an assembly, or the sheets of a drawing, most of the primary tools for editing, viewing, or selecting key elements of the design can be accessed with a right-click in the browser.

The Browser is located along the left edge of the Inventor window by default, but it can be relocated, made as a floating dialog, or even moved to another screen to maximize design window space.

Suppress and un-suppress part features

When editing, it can be very useful to temporarily Remove features from the model. One use for this is to correct errors that have been caused by changes to the model. By suppressing the feature that is failing, you can continue to make changes to the model until it's time to resolve the error.

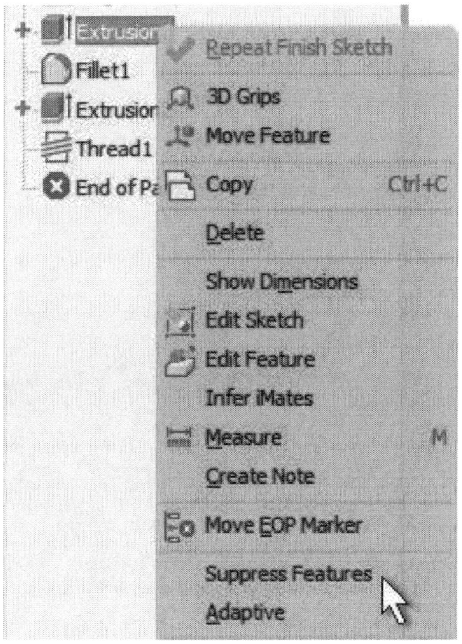

To suppress a feature located in the browser, right-click, and select suppress from the context menu. To restore the feature, use the same process to select un-suppress.

Components of the assembly can be suppressed as well. Suppressing a component in the context of an assembly removes it from memory. There are advanced assembly management techniques that use suppression of components to make even the largest assemblies manageable.

Toggle visibility of features and sketches

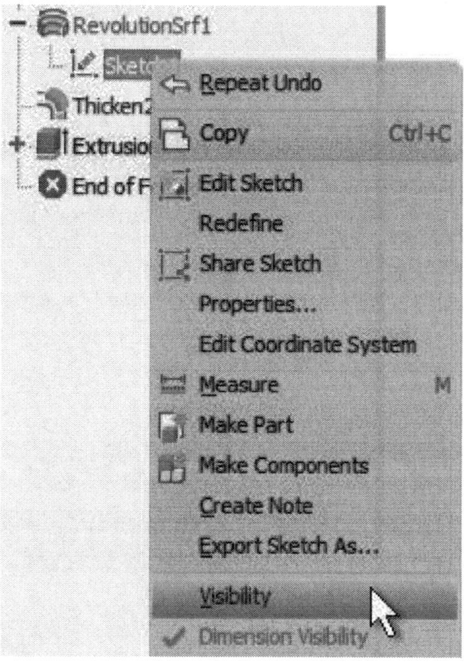

When a sketch has been used to create a feature, its visibility will automatically be turned off. At any time in the modeling process the sketch can be turned back on. One reason you might want to turn the visibility of the sketch back on is to use the sketch to create additional features. Using a visible sketch to create another feature will move it in the browser placing it before all other features. This is the same as sharing a sketch. Another reason is to be able to view the geometry or edit the dimension values that are applied to the sketch while working in the context of the component or assembly. This saves the step of activating the sketch to make a change.

The visibility of the sketch is controlled in the browser and can be toggled by selecting the sketch which is either located under a single feature or under the part with which it was shared and using right-click to access the context menu.

Chapter 8: Assembly Modeling Objectives

Most products regardless of their use are made up of more than one part. Defining the relationships between these parts or the preferred term, components of an assembly, is how you can verify how the design will fit, give guidance on how they will be assembled, and function in the real world.

Inventor uses a series of constraints to control the connections between components. Every ungrounded component has six degrees of freedom – three rotational and three translational. As you apply constraints fewer degrees of freedom will remain until you've removed as many as needed to replicate the relationships between components. Understanding what the assembly does and how you might want to modify it will guide your choice of constraints and making those choices can take time to master.

Ground base component of an assembly

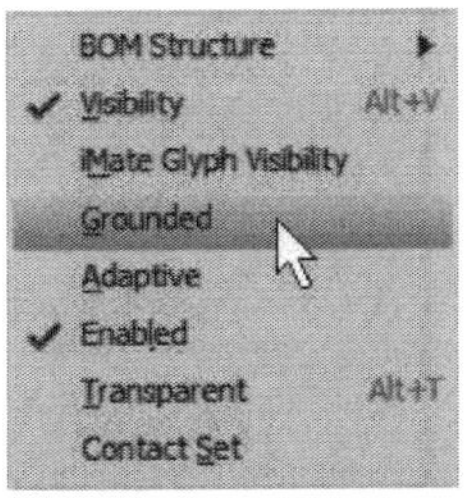

To declare a component is grounded is to say that it cannot move in relation to the coordinate system of the assembly. Often, the component that is placed in to the assembly first will be the part of the design that rests on the ground or is attached to another assembly.

Traditionally, people will ground this component to remove all degrees of freedom and begin constraining other components to it. Any number of components can be grounded in the assembly, and there are some advanced top down workflows where components are built in a specific position in their individual files and grounded in place in the assembly. You can ground a component through a right-click menu on the component in the design window or in the Browser and selecting Ground.

When inserting a component in the assembly, you can right-click and choose Place Grounded at Origin to match the position and orientation of the component's origin axes with the origin of the assembly.

Apply basic assembly constraints (mate, flush, insert, directed angle)

To position a component in an assembly and have it maintain its correct position, degrees of freedom must be removed. This is done using Constraint or Joint tools. Each of these tools has options that limit specific levels of freedom depending on what is selected.

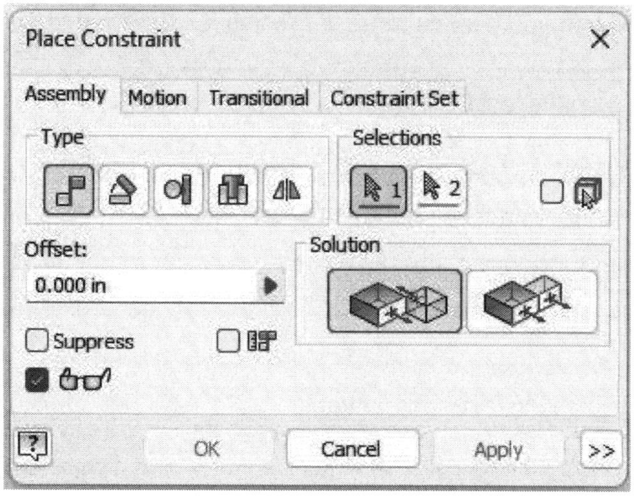

Constraints can rigidly connect components together or be used to construct the motion of a mechanism by removing only some of the degrees of freedom between components. These constraints can be applied to faces, edges, or construction geometry and different constraints are used for different scenarios. Some constraints can be used in combination with others to create relationships that can also be replicated using another constraint. Constraints can also have multiple ways of setting a direction or alignment referred to as solutions. These different solutions will allow the same constraint to position parts differently even if the same type of reference geometry is selected.

For reference, let's focus on a few of the most commonly used constraints and their most commonly used solutions. These tools can all be found in the Constraint tool on the Assemble tab.

The Mate constraint type is used to connect faces, edges, points or any combination of two entities. Selecting a face or edge will place an arrow icon along with the highlight. If you select two faces and use the Mate solution, the arrows on the selected entities will point toward or oppose each other. If the Flush solution is used, axes will not be selectable, and those arrows will go the same direction to "align" the selected faces.

An Angle constraint can use three different solutions to set an angular orientation between components. After selecting two entities which can be faces or edges that need to be set to a specific angle, you'll select a third reference which needs to be an edge, axis or curved face to define the axis which will act as the pivot point for the angle between the components. Once the three selections are made, you simply enter an angle value in the dialog and click OK. This value can be positive or negative. There are two additional solutions for the Angle constraint: Directed Angle and Undirected Angle. These options only need two selections to set the faces or edges that need to be positioned.

While using the Insert constraint is popular for placing bolts in holes, it can be used to make any curved faces with flat ends concentric and mated. It replaces the use of two Mate constraints to put the bottom of the bolt head on a face and mate the axis of the hole and the axis of the bolt. To use the insert constraint, you select a curved edge on each of the components. The planar faces the curves are on and the axes will be connected. One rotational degree of freedom will remain allowing the components to pivot on the common axis.

Apply an offset to constrained parts

Many of the constraints you apply have an Offset value in the dialog. Putting a positive value in the offset will create a gap between the components. You can also add a negative value to represent an interference. Adding a value will give you a preview of the new position of the components which will include the offset value. You can change the offset value after the constraint has been placed by selecting the constraint in the browser and entering a new value for the offset in the field that appears.

Determine the degrees of freedom of a component

To check your progress in adding assembly constraints you can view the degrees of freedom available on the components in the assembly by turning on the Degrees of Freedom display in the Visibility panel of the View tab. Pressing Ctrl + Shift + E will also turn on the display as a keyboard shortcut. Glyphs will display the number of remaining rotational and translational degrees of freedom. As constraints are added these glyphs will update to show what flexibility is still remaining in the assembly. The display of the glyphs can be switched on or off at any time.

Create a presentation model

The assembly data can be used for creating drawings, used in other assemblies, and it can also be used to create animations which can either showcase the design or to create exploded views or assembly animations.

In Autodesk Inventor, animations are created in a different file type, the presentation file which has an .ipn extension. To create a presentation file, you must first insert a linked Assembly file. This loads the current model geometry, appearance values, and its constraints.

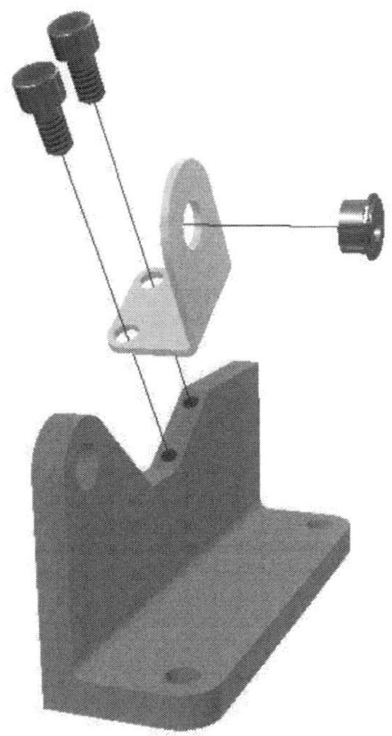

Once the assembly data is loaded, you can use the Tweak Components tool to move or rotate the components of the assembly to spread them apart. You can tweak as many or as few components as you want in individual events. These events are recorded in a storyboard and the timeline at the bottom of the screen. Moving the play head of the timeline allows you to create these events at different points in time and can be created with different durations to build a step by step animation.

You can also change the opacity of components at various steps on the timeline. Their visibility can be switched on or off immediately or the duration of the change can be stretched over time. The opacity events do not have to be aligned with any other event. Along with these changes, the view position of the model can be captured as a camera view to build motion of the view of the assembly as well as the tweaks and opacity. The animation will smoothly transition from camera position to camera position.

Once all of your animation looks like you want it in the storyboard you can export the animation as a video. You can also capture specific single views as a snapshot. A snapshot can be used to create a drawing view to be added to a drawing file, or to a raster image. You can create additional storyboards to represent different ways of displaying the model.

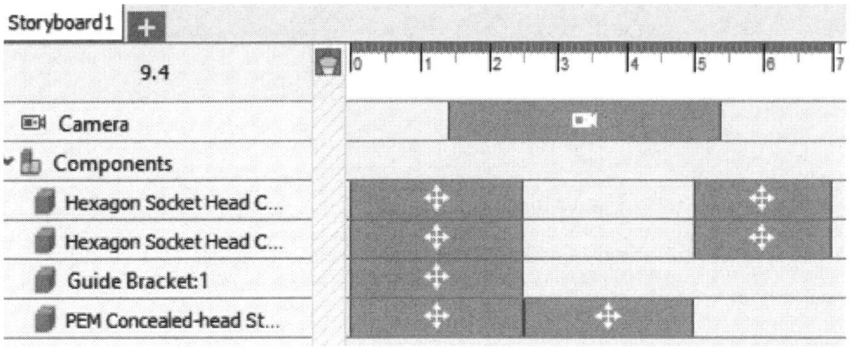

Notes:

Chapter 9: Drawing Objectives

With so much focus on 3D, it can be easy to forget 2D. Along with the ability to more easily and accurately check for fit and function in a 3D model, the ease of creating 2D detail drawings from the 3D model is one of the most compelling reasons people have switched to Inventor.

The certification objectives around drawings cover a broad array of tools and do a good job of covering the most important workflows.

Select and place a front view

To create a first or Base view on a drawing select the Base view tool on the Place Views tab. By default, Inventor will present a preview of the view after you have selected a part or assembly to detail. If you create a new drawing with only one 3D file open, Inventor will select it automatically. You can click and drag the preview to another location on the sheet if needed. The alignment of the ViewCube in the Inventor model determines what view of the model will be offered as the Front view in the drawing.

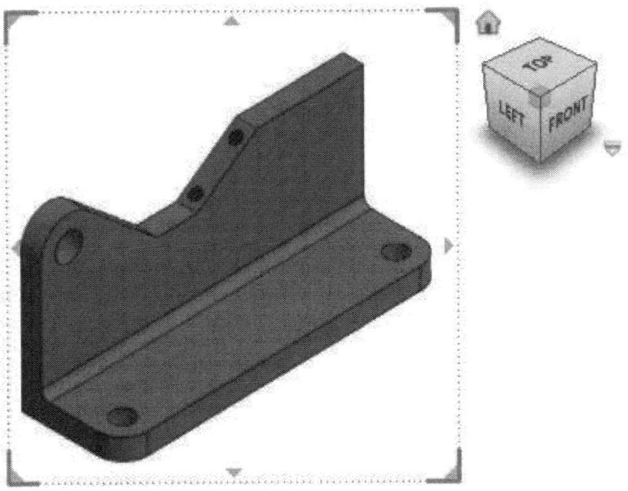

A graphic of the ViewCube will be shown near the view preview allowing you to choose what view the Base view will be created from. You can also rotate the view orientation, choose the Home View of the model or choose any corner or edge of the ViewCube to position the drawing view.

A bounding box preview will appear if you move your cursor away from the view preview. If you click to locate a preview, it will generate an orthographic or isometric child view along with the Base view. Another option for placing orthographic projected views along with the Base view is to click the arrows that appear at the top, bottom, and sides of the view. Selecting additional view placements will generate a preview of those views as well. Any change to the position, view point, or scale of the Base view will update the orthographic projections as well.

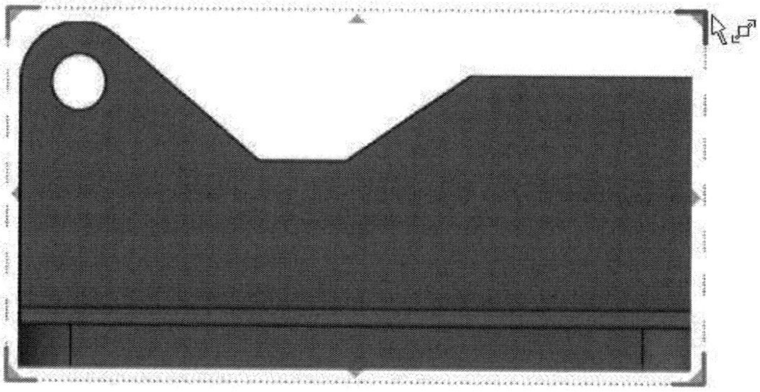

The preview will automatically be scaled based on the drawing sheet size and the model size. Highlighted corners on the view preview make it easy to click and drag the view scale to be larger or smaller to fit the sheet as you need. Any changes to the scale can be made in the Drawing View dialog that appears.

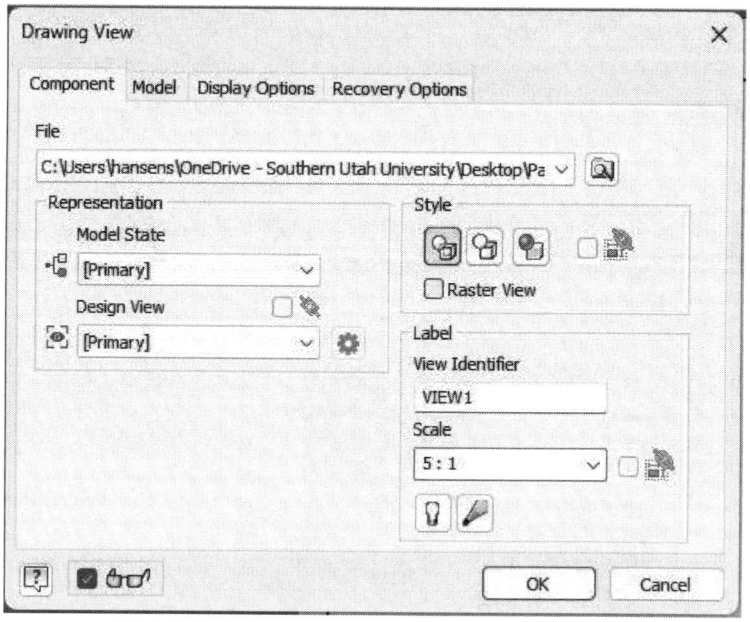

This dialog can also be used to change additional properties of the drawing view. Options within the dialog give you the ability to choose what representations will be used, if Hidden lines will be generated, if the view will be shaded, and many other more advanced options.

Once you have the orientation and position set to your needs, you will need to click OK to create the view or views as you placed them on the sheet.

Create a drawing view from an existing view

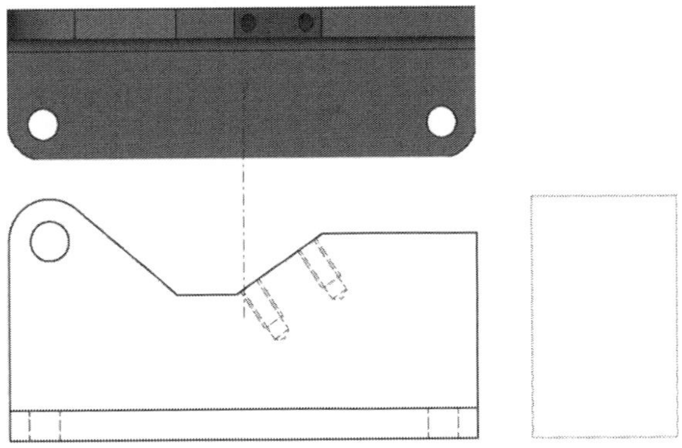

Multiple views can be created along with the initial Base view, but you can add projected views based on virtually any other view whenever you need. Starting the Projected View tool, you will be prompted to select a parent view.

The view you'll be creating, referred to as a child view, will get its scale and view style from the parent view and will have its position related to it so all you will need to do is move the preview into position, right-click, and choose Create from the marking menu to generate the drawing view.

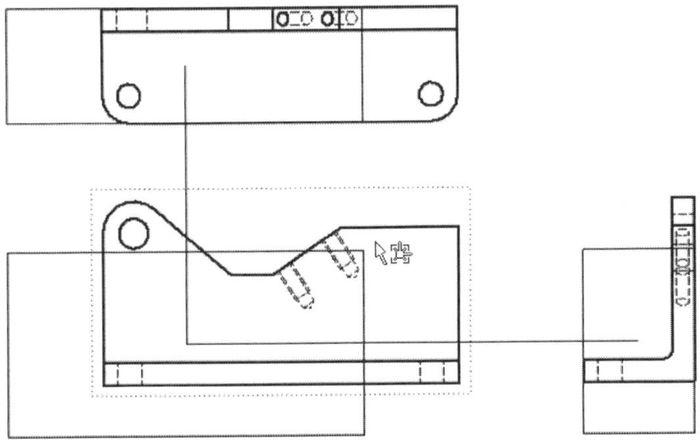

Once the view is placed any movement or change in style or scale of the parent view will automatically update the child view. If you would like to change the properties of the child view, some limited options are available. With a double-click on the drawing view the Drawing View dialog will open and you'll be able to deselect the linking of the view style and scale from the parent view. At any point you can choose to re-associate the style and scale to the parent if you need.

Add annotation and dimensioning to a drawing

More often than not the value of a 2D detail drawing is derived from Dimensions applied to the drawing views. These dimensions inform the viewer of the critical sizes that a component must be made to, the overall size of an assembly, or the range of motion of a mechanism. When these dimensions are applied, they derive their value for the geometry of the model and are not affected by the scale of the particular drawing view you are applying dimensions to.

Dimensions are placed by selecting geometry in the drawing View. Using the Dimension tool, the type of geometry you select will offer different dimension types. Selecting a single line will give you a Linear dimension. If the line selected is not vertical or horizontal, moving the mouse around near the view might offer you a horizontal, vertical, or aligned dimension. You can right-click to force an option under the Dimension Type fly-out. You can also place linear dimensions by selecting parallel edges, two points or a point and an edge.

Once the geometry is selected, you move the cursor to where you want to place the dimension and click to place it. The Dimension tool will continue to be ready to place dimensions of any type until you press the Esc key or select Cancel from the Marking Menu.

Selecting two non-parallel lines will cause the Dimension tool to place an angular dimension. Once the geometry is selected, as you move around you can see the angular dimension options for that geometry.

The Dimension tool will also place radial and diameter dimensions based on selecting an arc or a circle. You can again switch between radial and diameter using a right-click option. Clicking on a straight edge that represents a circular face will offer a linear diameter dimension.

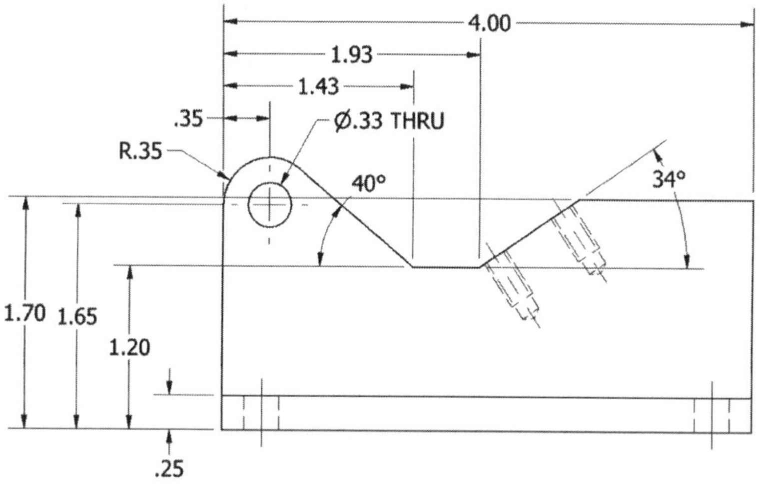

There are additional dimensioning tools for placing groups of dimensions or dimensions that build on each other and are placed at once after selecting multiple points. The baseline dimension tools place collections of dimensions off a common datum point. Chain dimensions connect dimensions end to end in a row after selecting the points.

Other dimension types and annotation types can be added to drawing views as well. If you need ordinate dimensions, which are commonly used for machined components, the process is very much like placing chain dimensions. You can add leaders with text or symbols and add specialized callouts for finish symbols and geometric dimensioning and tolerance (GDT) symbology and definition. When adding Hole Notes, the information on the hole will be added to the dialog directly from the hole feature itself. You still can modify the callout for style as needed through the dialog.

Create a drawing view based on an assembly and presentation file

The process for creating drawing views of any part, assembly, or presentation file is essentially the same. There are some differences in the defaults. Assemblies will remove hidden lines from the view by default for clarity assuming that the internal details of the components will be called out in the drawings of the individual parts.

When creating a view of a Presentation file, you will need to choose which snapshot will be used to create the drawing view. Snapshots can be created in any of the presentation file's storyboards so they can offer different viewpoints, separation values and combinations of component visibility.

An individual part within an assembly can have its visibility turned off, have hidden lines displayed, can be made transparent, or can have its participation in any section views managed. This is true in views of an assembly or presentation file and these options do not depend on the visual properties of the components in the source assembly file.

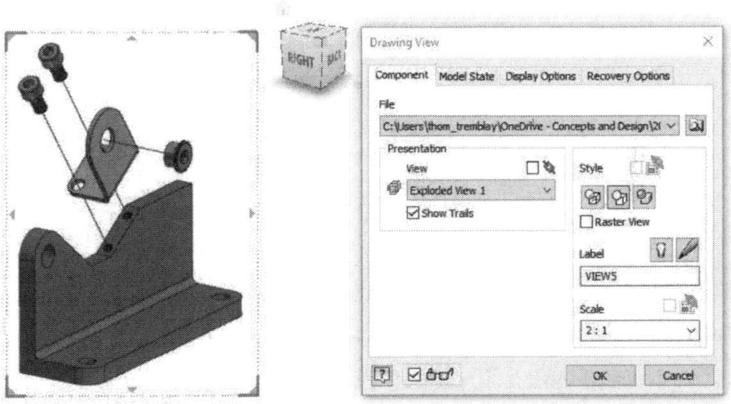

Add balloons to a drawing

It is common practice to add balloon callouts to assembly and exploded views made from presentation files. It makes it much easier to associate the part to a parts list that can exist on the active drawing sheet or a drawing sheet located on another page.

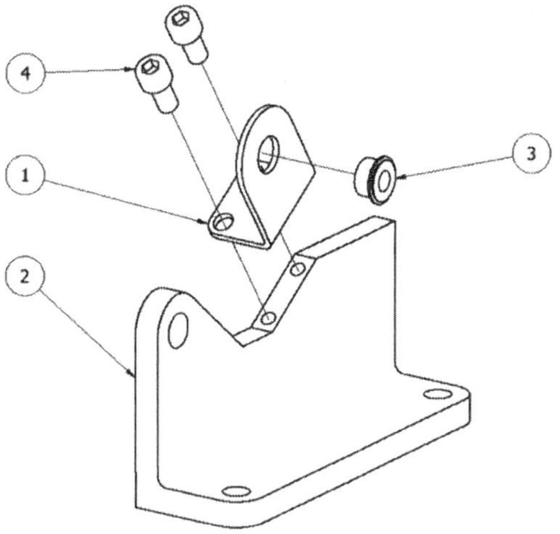

Balloons can be placed individually using the Balloon tool on the Table panel of the Annotation tab or multiple balloons can be placed at once using Auto Balloon.

The description in the balloon, typically a number, is associated to the item number in the parts list. It can be overridden or changed to another type of displayed value. The item number can be changed from the balloon or the parts list.

Create and edit a parts list in a drawing

A parts list is a table that represents the bill of materials which is generated and maintained as you add or remove components from the assembly. The parts list is displayed as a table on the drawing. Its format can be edited to change column spacing, appearance, etc. The content can also be edited to override the values generated by the bill of materials.

PARTS LIST			
ITEM	QTY	PART NUMBER	DESCRIPTION
1	1	Guide Bracket	
2	1	Support Bracket	
3	1	CSS-0420-3	PEM Concealed-head Standoffs CSS - Inch
4	2	ANSI B18.3 - No. 10 - 24 UNC - 3/8 HS HCS	Hexagon Socket Head Cap Screw

Normally, changes to the structure of how components are in the parts list, are made in the bill of materials and reflected in the parts list. However, you may choose to change item numbers or add additional description information about the components from the parts list. Additions made to the parts list such as description information will not be reflected in the bill of materials. If you want the additional description information connected to the component, you need to change the bill of materials.

Add sheets to a drawing

A drawing sheet can have any number of views that you need to fit on the page. There might be times where you need multiple sheets for additional drawing views. A drawing file can have any number of sheets for developing detail drawing views.

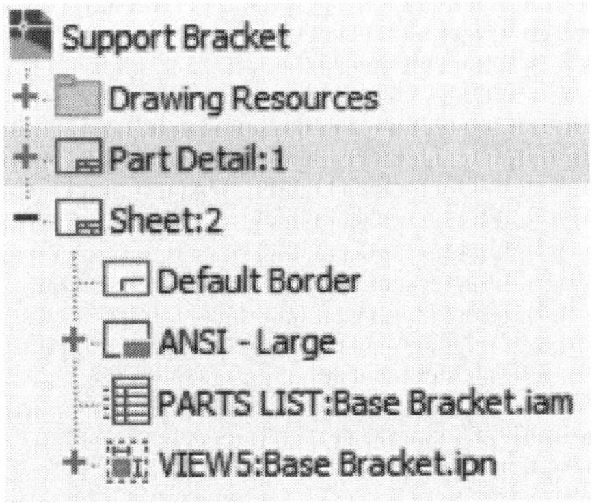

Adding additional sheets can be done using the New Sheet tool in the Place Views tab or by selecting the tool from the context menu when you right click on the drawing file header in the browser.

Any sheet of the drawing can have its own values for sheet size, and what, if any, border or Title Block it uses. This makes it easy to create drawings of an assembly and its various parts in a single drawing file if you should choose to do so.

Control sheet size and add a title block

Each sheet added to a drawing is displayed in the Browser on the left. If you expand the sheet you will see the elements border, title block, drawing views and the files displayed in the drawing views.

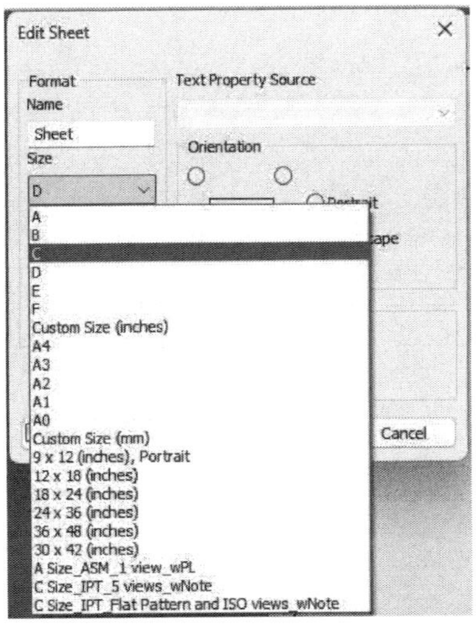

Changing the size of a drawing sheet requires you to edit the sheet using a right-click on the sheet in the browser. Once you select Edit Sheet you can change the name, sheet size and other variables in the dialog that is presented. You might find that you need to change the scale of the drawing views or reposition them on the updated sheet, but that is a simple process.

The title block can be different from sheet to sheet or a sheet can even have the title block removed. To edit or delete which title block is used or even edit the format of the title block, you must have the sheet active in the browser. Once it is active, a right-click on the title block in the browser will give you the options you need to make the change.

Notes:

Chapter 10: Practice Exam

In addition to the quiz questions in the Appendix, this book offers you access to an online exam to test your skills in a simulated Certification Exam environment. The purpose of the exam is to give you experience in working with Inventor in a timed examination environment so you can get used to how you would approach the real exam and how to work between the exam application and Inventor.

Before taking the Practice Exam

To take the exam you must take the following steps:

- You must have Autodesk Inventor 2025 installed on your system.
- Download the dataset and expand it on your computer's hard drive.
- Set your Inventor Project file to the *Fishing Rod.ipj* from the dataset folder.
- Download the Practice Exam software from SDC Publications using the instructions on the inside cover of this book.

About the practice exam

The exam is a collection of thirty random questions addressing each of the certification objective domains. The questions reflect the flow of the actual exam by asking you to perform tasks with often limited information on the steps you will need to take to execute them.

These are not the questions in the actual exam or the identical solution workflows. Most of the answers to the questions will be contained in the material properties of the model after editing, so efficient use of time and understanding how to use the Properties dialog is essential.

The minimum passing grade for the practice exam is 70%. You may take the exam more than once, but you should be sure to wait until you've read the book and reviewed the quiz questions and answers to seek out areas where you might need to do more study and gain more hands-on experience before taking the practice exam.

Preparing for the practice exam

The exam will require you to download two sets of data, the exam files and the datasets for the exam. The dataset is a sample dataset that is available from:

https://knowledge.autodesk.com/sites/default/files/file_downloads/Fishing_Rod.zip

Once it is finished downloading, extract the *Fishing Rod* folder from the .zip file to a location that is easy for you to find on your computer.

The exam itself is contained in an html page that will be launched from your computer. By following the instructions on the inside front cover page of this book you will be able to download a .zip file which contains the files that you need.

Once you've downloaded the .zip file, extract the contents to a folder on your computer that you can remember.

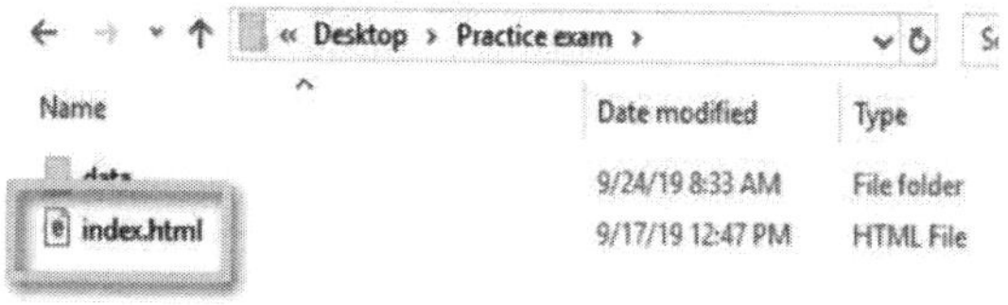

Starting the practice exam

Before attempting to start the exam, launch Autodesk Inventor and set the project file to the Fishing Rod.ipj. Use the Open tool to make sure that the Fishing Rod files are easily accessible. Once you're sure the files are where they need to be, minimize Inventor. Having Inventor running before beginning the official Certification exam is necessary, so this will get you accustomed to the action. Now find the folder with the extracted exam files, open the folder and double-click the *index.html* file to open the exam in a web browser window. Once the browser window with the exam opens, you'll be prompted to click **Start Quiz** to begin.

This will open a new page where you have the option of entering your name and e-mail address. Entering this information will make it possible for the test to send you your results in a personalized message. Click the **Submit** button in the bottom-right to move to the next page.

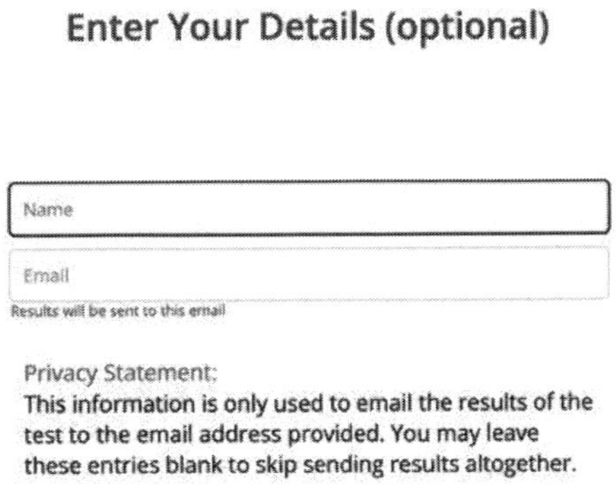

The next page will offer critical information on the formatting of the answers you'll submit.

- Answers must include all digits.
- Units must be included in the answer.
- Do not create a space between numeric values and units call out.

Click **Continue** to begin the 50-minute timer and present the first question of the exam.

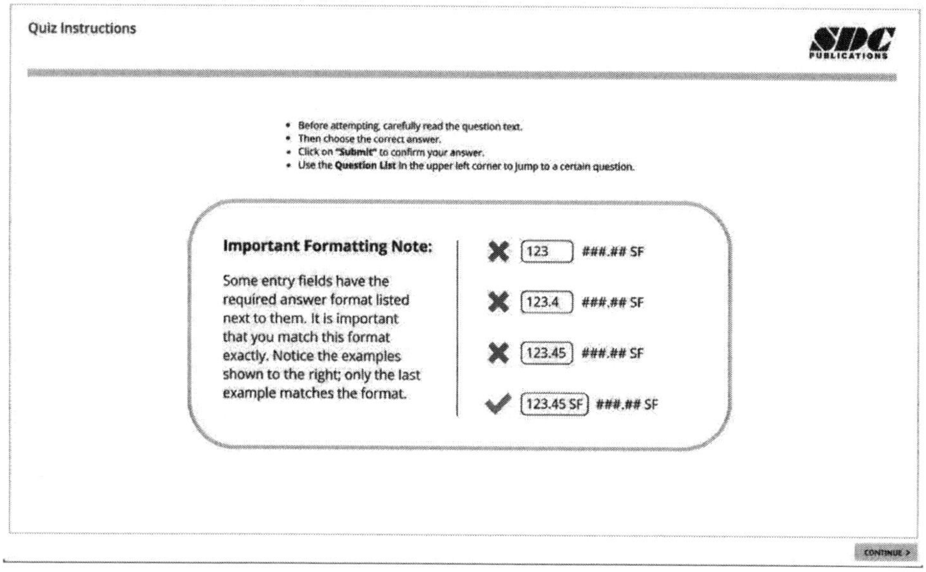

You will be offered 30 questions. The information on what, if any, file that you will need to open will be at the top of the screen. Below the gold bar on the screen will be the question and any information on the task that needs to be performed. At the end of the question will be the answer field which is blank space where you will enter your solution that you find.

Note: Remember, for questions that include units, be sure to include the unit as displayed next to the answer field.

After you enter a value, click **Submit** to have the question graded. If you enter an incorrect value, you will receive a notice at the bottom with a page reference for you to review the certification objective and improve your ability to pass the exam.

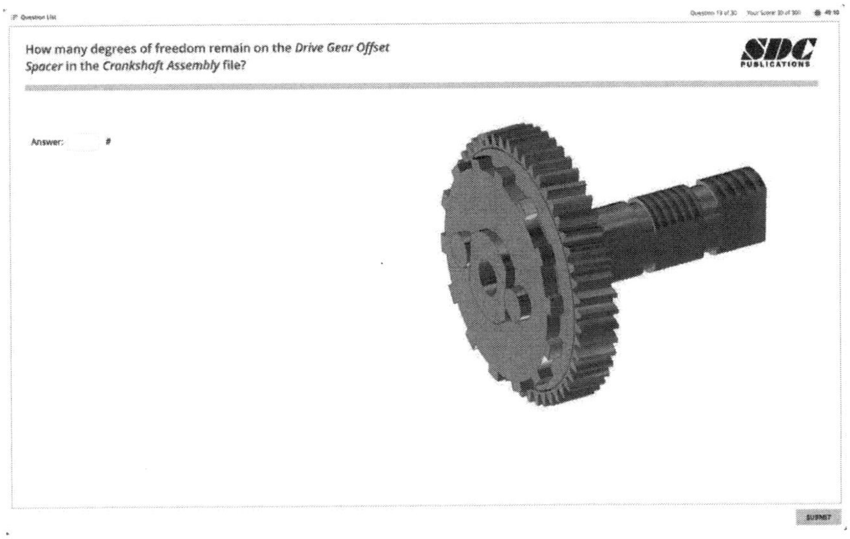

Repeat the process of opening files when asked to and executing the tasks. If you encounter a question that you're unsure of how to execute, mark the question and move on. You can go back to any question or skip forward by selecting the Question list icon in the upper left corner and then choosing the question from the list that appears.

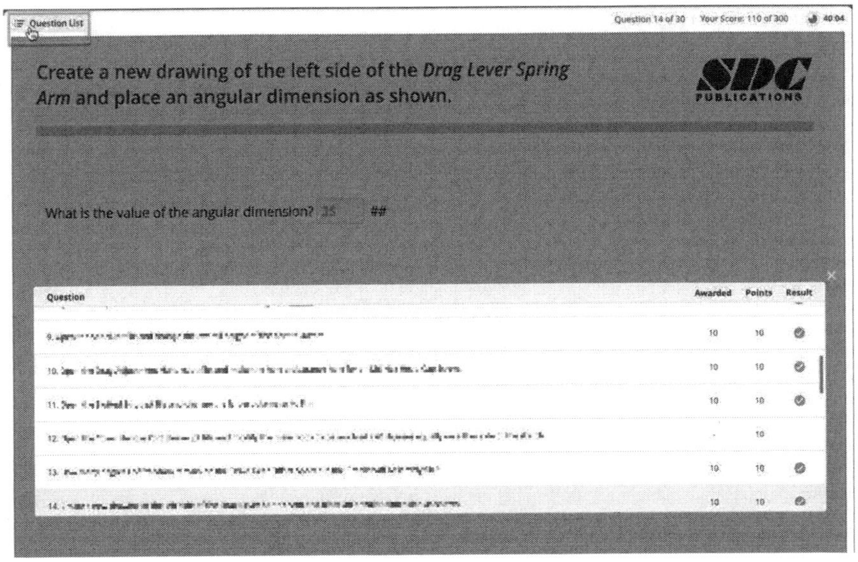

When you've completed all of the questions or if time expires, the button in the lower right will change to read "View Results". Clicking the button will display whether you passed or not and what your score was against the passing score.

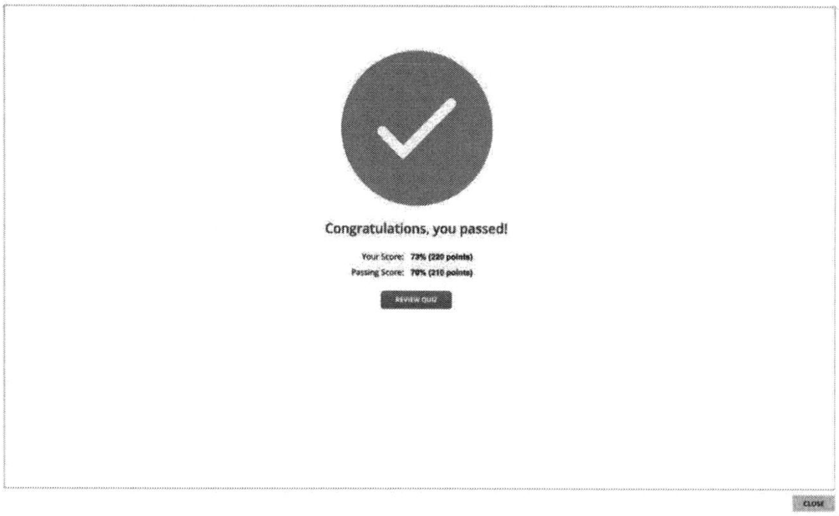

You can review the quiz that you just completed and use the **Prev** and **Next** buttons to walk through the individual questions and review your results. This is a perfect opportunity to make note of the page numbers if you were too focused to do so during the timed exam.

When you're finished with your review, click the **Close Review** button in the lower left to return back to the results page where you can click the **Close** button to exit the exam.

Your results

If you entered your e-mail information you will receive a message with your results.

The e-mail will also include a listing of the questions, your answers and the pages in the book you can refer to for study information.

You can take the exam again to work on imporoving your score or to just build up your skills for taking the certification exam.

Summary

The practice exam offers a tremendous opportunity to evaluate your skills in a timed environment executing problems and the examination process that resembles the process used for the official exam without additional expense and in a location of your choice.

Best of luck on the practice exam and on becoming an Autodesk Inventor Certified User.

Notes:

Appendix A: Practice Test

The following questions will be related to Chapters 4 through 9 and limited to topics found in the certification objective domains. These questions are intended to give you an opportunity to review the type of question that might be asked during the Autodesk Inventor User Certification exam. Some of these same objective domains will appear in the professional exam so there might be overlap, but it would be a coincidence.

It is advisable to use the Autodesk Inventor software to help you locate the answer and even if you are confident that you know the correct answer to verify it before choosing an option.

To get the best results for evaluating your preparedness, complete the following questions in fifty 50 minutes or less.

1. **The viewpoint of a model can always be restored to its original position using:**

☐ Shift Key

☐ Home View

☐ Restore View

☐ Mouse wheel double-click

2. **How many dimensioning tools are typically used in creating a sketch?**

☐ 1

☐ 3

☐ 5

☐ 7

3. **Autodesk Inventor assemblies use the _____ file format.**

☐ .ipn

☐ .ipt

☐ .dwg

☐ .iam

4. A one step constraint for placing a mating relationship on a face and axis is:

☐ Insert

☐ Bolted

☐ Flush

☐ Align

5. To duplicate a feature around an axis, use the ____ tool:

☐ Circular pattern

☐ Rectangular pattern

☐ Mirror feature

☐ Round array

6. A projection of an existing view is referred to as a ____ view.

☐ Child

☐ Section

☐ Cast

☐ Offset

7. When applying Colinear constraint, selected entities will automatically be made Horizontal.

☐ True

☐ False

8. A feature commonly used to hollow out a part is the ____ feature.

☐ Thin wall

☐ Fillet

☐ Empty

☐ Shell

9. Sketch geometry that is used to define a sketch but not create 3-D geometry is called:

☐ Projected geometry

☐ Construction geometry

☐ Offset entities

☐ A closed loop

10. The first view placed in a drawing is the ____ view regardless of its orientation.

☐ Auxiliary

☐ Start

☐ Base

☐ Projected

11. Feature that creates a rounded face on an edge is called a:

☐ Revolve feature

☐ Fillet feature

☐ Sweep feature

☐ Extrude feature

12. Geometry projected into a sketch will _______ by default.

☐ Be created on a separate layer

☐ Extrude their shape

☐ Update with changes to the source geometry

☐ Have a hidden linetype

13. How many sheets can be in a drawing?

☐ 1

☐ 5

☐ 10

☐ More than 10

14. The guiding principle for Autodesk Inventor sketches is:

☐ Area loop

☐ Sectional

☐ Parametric

☐ Isometric

15. Autodesk Inventor can create 2-D drawings in DWG format directly.

☐ True

☐ False

16. Feature that creates an angled face on an edge is called a(n):

☐ Revolve feature

☐ Fillet feature

☐ Chamfer feature

☐ Extrude feature

17. An Extrude feature can only be created in one direction.

☐ True

☐ False

18. Hole features cannot automatically be sized based on fastener.

☐ True

☐ False

19. Removing a feature from the part temporarily is called ____ the feature.

☐ Freezing

☐ Suppressing

☐ Isolating

☐ Disabling

20. To make a component no longer appear in an assembly without removing it from memory use:

☐ Disable

☐ Hide

☐ Off

☐ Visibility

21. Feature that builds a 3-D shape around an axis based on a sketch is called a(n):

☐ Revolve feature

☐ Fillet feature

☐ Sweep feature

☐ Extrude feature

22. Removing all degrees of freedom without placing a constraint is making a component:

- ☐ Fixed
- ☐ Free
- ☐ Grounded
- ☐ Locked

23. How many solution types does the Mate constraint have?

- ☐ 1
- ☐ 2
- ☐ 4
- ☐ 5

24. A Presentation file uses the ____ file extension.

- ☐ .iam
- ☐ .ipt
- ☐ .ipn
- ☐ .idw

25. Most detail dimensions can be placed in a drawing view using one Dimension tool.

- ☐ True
- ☐ False

26. A graphical representation of events in an animation is referred to as a(n):

- ☐ Slate
- ☐ Event track
- ☐ Timeline
- ☐ Storyboard

27. A part file in Inventor uses the ____ file format.

- ☐ .ipn
- ☐ .ipt
- ☐ .dwg
- ☐ .iam

28. Feature that builds a 3-D shape normal to a sketch is called a(n):

- ☐ Revolve feature
- ☐ Fillet feature
- ☐ Chamfer feature
- ☐ Extrude feature

29. No offset space can be defined in a Constraint.

- ☐ True
- ☐ False

30. A Balloon placed has a number added to it based on the _____ by default.

- ☐ Part size
- ☐ Material
- ☐ Item number
- ☐ Quantity

Appendix B: Practice Test Answers

The following questions will be related to Chapters 4 through 9 and limited to topics found in the certification objective domains. These questions are intended to give you an opportunity to review the type of question that might be asked during the Autodesk Inventor User Certification exam. Some of these same objective domains will appear in the professional exam so there might be overlap, but it would be a coincidence.

It is advisable to use the Autodesk Inventor software to help you locate the answer and even if you are confident that you know the correct answer to verify it before choosing an option.

To get the best results for evaluating your preparedness, complete the following questions in fifty 50 minutes or less.

1. The viewpoint of a model can always be restored to its original position using:

- ☐ Shift Key
- ■ **Home View**
- ☐ Restore View
- ☐ Mouse wheel double-click

2. How many dimensioning tools are typically used in creating a sketch?

- ■ **1**
- ☐ 3
- ☐ 5
- ☐ 7

3. Autodesk Inventor assemblies use the _____ file format.

- ☐ .ipn
- ☐ .ipt
- ☐ .dwg
- ■ **.iam.**

4. A one step constraint for placing a mating relationship on a face and axis is:

- ■ **Insert**
- ☐ Bolted
- ☐ Flush
- ☐ Align

5. To duplicate a feature around an axis, use the ____ tool:

- ■ **Circular pattern**
- □ Rectangular pattern
- □ Mirror feature
- □ Round array

7. When applying Colinear constraint, selected entities will automatically be made Horizontal.

- □ True
- ■ False

9. Sketch geometry that is used to define a sketch but not create 3-D geometry is called:

- □ Projected geometry
- ■ Construction **geometry**
- □ Offset entities
- □ A closed loop

11. Feature that creates a rounded face on an edge is called a:

- □ Revolve feature
- ■ Fillet **feature**
- □ Sweep feature
- □ Extrude feature

13. How many sheets can be in a drawing?

- □ 1
- □ 5
- □ 10
- ■ More **than 10**

6. A projection of an existing view is referred to as a ____ view.

- ■ **Child**
- □ Section
- □ Cast
- □ Offset

8. A feature commonly used to hollow out a part is the ____ feature.

- □ Thin wall
- □ Fillet
- □ Empty
- ■ **Shell**

10. The first view placed in a drawing is the ____ view regardless of its orientation.

- □ Auxiliary
- □ Start
- ■ **Base**
- □ Projected

12. Geometry projected into a sketch will ________ by default

- □ Be created on a separate layer
- □ Extrude their shape
- ■ **Update the source geometry**
- □ Have a hidden linetype

14. The guiding principle for Autodesk Inventor sketches is:

- □ Area loop
- □ Sectional
- ■ **Parametric**
- □ Isometric

15. Autodesk Inventor can create 2-D drawings in DWG format directly.

- ■ **True**
- □ False

16. Feature that creates an angled face on an edge is called a(n):

- □ Revolve feature
- □ Fillet feature
- ■ **Chamfer feature**
- □ Extrude feature

17. An Extrude feature can only be created in one direction.

- □ True
- ■ **False**

18. Hole features cannot automatically be sized based on fastener.

- □ True
- ■ **False**

19. Removing a feature from the part temporarily is called ____ the feature.

- □ Freezing
- ■ **Suppressing**
- □ Isolating
- □ Disabling

20. To make a component no longer appear in an assembly without removing it from memory use:

- □ Disable
- □ Hide
- □ Off
- ■ **Visibility**

21. Feature that builds a 3-D shape around an axis based on a sketch is called a(n):

- ■ **Revolve feature**
- □ Fillet feature
- □ Sweep feature
- □ Extrude feature

22. Removing all degrees of freedom without placing a constraint is making a component:

- □ Fixed
- □ Free
- ■ **Grounded**
- □ Locked

23. How many solution types does the Mate constraint have?

- □ 1
- ■ **2**
- □ 4
- □ 5

24. A Presentation file uses the ____ file extension.

- □ .iam
- □ .ipt
- ■ **.ipn**
- □ .idw

25. **Most detail dimensions can be placed in a drawing view using one Dimension tool.**

- ■ **True**
- □ False

26. **A graphical representation of events in an animation is referred to as a(n):**

- □ Slate
- □ Event track
- □ Timeline
- ■ **Storyboard**

27. **A part file in Inventor uses the ____ file format.**

- □ .ipn
- ■ **.ipt**
- □ .dwg
- □ .iam

28. **Feature that builds a 3-D shape normal to a sketch is called a(n):**

- □ Revolve feature
- □ Fillet feature
- □ Chamfer feature
- ■ **Extrude feature**

29. **No offset space can be defined in a Constraint.**

- □ True
- ■ **False**

30. **A Balloon placed has a number added to it based on the _____ by default.**

- □ Part size
- □ Material
- ■ **Item number**
- □ Quantity

PGIL2026USA